I0470659

THIS PAGE:
Plain Run Branch, in Appomattox Court House National Historical Park, is one tributary of the Appomattox River (also present in the park). The boulders are Cambrian-age metamorphic rock that formed deep within the Earth, in association with the Appalachian Mountains.

ON THE COVER:
The rolling forested hills of the Piedmont Province of Virginia form the landscape of Appomattox Court House National Historical Park. The low hills contain rocks that formed deep within the Earth during the uplift of the Appalachian Mountains.

NPS Photos courtesy Brian Eick (NPS APCO)

Appomattox Court House National Historical Park

Geologic Resources Inventory Report

Natural Resource Report NPS/NRPC/GRD/NRR—2009/145

Geologic Resources Division
Natural Resource Program Center
P.O. Box 25287
Denver, Colorado 80225

September 2009

U.S. Department of the Interior
National Park Service
Natural Resource Program Center
Denver, Colorado

The National Park Service, Natural Resource Program Center publishes a range of reports that address natural resource topics of interest and applicability to a broad audience in the National Park Service and others in natural resource management, including scientists, conservation and environmental constituencies, and the public.

The Natural Resource Report Series is used to disseminate high-priority, current natural resource management information with managerial application. The series targets a general, diverse audience, and may contain NPS policy considerations or address sensitive issues of management applicability.

All manuscripts in the series receive the appropriate level of peer review to ensure that the information is scientifically credible, technically accurate, appropriately written for the intended audience, and designed and published in a professional manner. This report received informal peer review by subject-matter experts who were not directly involved in the collection, analysis, or reporting of the data.

Views, statements, findings, conclusions, recommendations, and data in this report are those of the author(s) and do not necessarily reflect views and policies of the National Park Service, U.S. Department of the Interior. Mention of trade names or commercial products does not constitute endorsement or recommendation for use by the National Park Service.

Printed copies of reports in these series may be produced in a limited quantity and they are only available as long as the supply lasts. This report is also available online from the Geologic Resources Inventory website (http://www.nature.nps.gov/geology/inventory/gre_publications.cfm) and the Natural Resource Publication Management website (http://www.nature.nps.gov/publications/NRPM/index.cfm) or by sending a request to the address on the back cover.

Please cite this publication as:

Thornberry-Ehrlich, T. 2009. Appomattox Court House National Historical Park Geologic Resources Inventory Report. Natural Resource Report NPS/NRPC/GRD/NRR—2009/145. National Park Service, Denver, Colorado.

NPS 340/100277, September 2009

Contents

Figures iv

Executive Summary v

Introduction 1
    Purpose of the Geologic Resources Inventory 1
    Park Setting 1

Geologic Issues 5
    Surface Water Issues and Channel Morphology 5
    Seismicity and Mass Wasting 6
    Anthropogenic Impacts 7
    Mine Features 7
    Hydrogeologic Modeling 7
    Future Facility Planning 8
    Miscellaneous Action Items 8

Geologic Features and Processes 11
    Geologic Structures 11
    Geology, Biology, and History Connections 11

Map Unit Properties 14
    Map Unit Discussion 14

Geologic History 19
    Proterozoic Eon 19
    Late Proterozoic Eon–Early Paleozoic Era 19
    Middle Paleozoic Era 20
    Late Paleozoic Era 20
    Mesozoic Era 21
    Cenozoic Era 21

Glossary 28

References 32

Appendix A: Geologic Map Graphic 35

Appendix B: Scoping Summary... 37

Attachment 1: Geologic Resources Inventory Products CD

Figures

Figure 1. Maps of Appomattox Court House National Historical Park vi
Figure 2. Location of the park relative to the physiographic provinces and subprovinces of Virginia 3
Figure 3. Locaction of the park relative to the geologic terranes of the Piedmont physiographic province of Virginia ... 4
Figure 4. The "Tibbs vernal pool" in Appomattox Court House National Historical Park 9
Figure 5. McLean House, the surrender site. Photo by Landry C. Thornberry. 10
Figure 6. Appomattox County Courthouse. Photo by Landry C. Thornberry. 10
Figure 7. Generalized geologic map of Appomattox and South Boston 30- x 60-minute quadrangles 13
Figure 8. Geologic time scale 23
Figure 9. Geologic time scale specific to Virginia 24
Figure 10. Geologic evolution of the Appalachian Mountains in the Appomattox area (Proterozoic–Cambrian) 25
Figure 11. Geologic evolution of the Appalachian Mountains in the Appomattox area (Cambrian–Mississippian) 26
Figure 12. Geologic evolution of the Appalachian Mountains in the Appomattox area (Permian–Present) 27

Executive Summary

This report accompanies the digital geologic map for Appomattox Court House National Historical Park in Virginia, which the Geologic Resources Division produced in collaboration with its partners. It contains information relevant to resource management and scientific research. This document incorporates preexisting geologic information and does not include new data or additional fieldwork.

Established in 1930, Appomattox Court House National Historical Park covers 705 ha (1,743 ac) in Appomattox County, Virginia, and preserves the village of Appomattox Court House and the surrounding farm, forest, and meadow lands. The purpose of the park is to commemorate the surrender of the Confederate Army of Northern Virginia to the Union Army on April 9, 1865, the strategic act that effectively ended the American Civil War. The park preserves many cultural resources, including the site of surrender at McLean House, the historic village at Appomattox, and the site of the Battle of Appomattox Court House.

The park lies within the western Piedmont physiographic province, which is underlain by hard crystalline igneous and metamorphic rocks such as schists, phyllites, slates, gneisses, and gabbros. This portion of the Piedmont physiographic province contains numerous belts of rocks that accreted onto the North American continent during repeated continental collisions and orogenic events, culminating in the formation of the Appalachian Mountains. This geologic history spans more than 700 million years through the present regime of weathering, erosion, and anthropogenic land use patterns. Erosion has dissected this terrain into gently rolling hills and valleys.

Because geology forms the foundation of the ecosystem and has influenced historical events in the Appomattox area, geologic issues such as surface water quality, channel morphology, seismicity, mass wasting, mine features, and hydrogeology are significant to the park's resource managers. Geologic knowledge is also essential in understanding landscape evolution and anthropogenic impacts, and for the siting of future facilities.

Appomattox Court House National Historical Park protects some of the headwater region of the Appomattox River. For this reason, the preservation and protection of surface water and groundwater resources at the park is vital for the greater ecosystem. State and federal Clean Water Act water quality standards are currently being met within the park.

Geologic processes such as erosion can impact the cultural landscape and natural systems in the park. Ground shaking associated with seismicity can damage historic structures. Mass wasting distorts the landscape and can increase sediment loads in local waterways, damaging the aquatic environment. Several small-scale topographic depressions exist within the park boundaries, but the exact nature of these features is unknown. Resource managers need to understand how water is moving through and under the park; thus, hydrogeologic modeling is critical in determining the impacts of human-induced contaminants on the entire ecosystem as well as in predicting groundwater table changes.

Rocks in the Appomattox Court House area typically occur in elongated parallel belts trending northeast-southwest and dipping steeply to the southeast. This orientation reflects the mountain-building processes that occurred during the accretion of terranes in the Paleozoic Era (Nelson et al. 1999; Hackley et al. 2007). The rocks of this area reflect a complex deformational and metamorphic history. The park sits atop Proterozoic ("Precambrian")- to Cambrian-age bedrock that contains metamorphic rocks such as mica schist, biotite gneiss, amphibolite, and a band of greenstone and amphibole gneiss, with an interlayered belt of silica-rich metamorphosed tuff, mica schist, and gneiss. Cutting through these rocks are small-scale faults, fractures, mylonite zones, foliation, and other deformation features; small-scale joints and fractures are ubiquitous. Most joints and fractures follow the northeast-southwest regional structural trend. Parallel thrust faults cut through the Appomattox County area, and may still be accommodating crustal stresses.

From the first settlers and farmers through modern development, geology has played a fundamental role in the evolution of Appomattox Court House National Historical Park. The area's geology is more than just a collection of rocks and mineral resources; geologic features and processes affect topographic expression, streams and rivers, soils, wetlands and bogs, vegetation patterns, and the animal life that thrives in the environment.

The geological landscape in south-central Virginia underscores the unique relationship between geology and history. Both the geologic resources and the natural history of the area should be emphasized and interpreted to enhance the visitor's experience of Appomattox Court House National Historical Park.

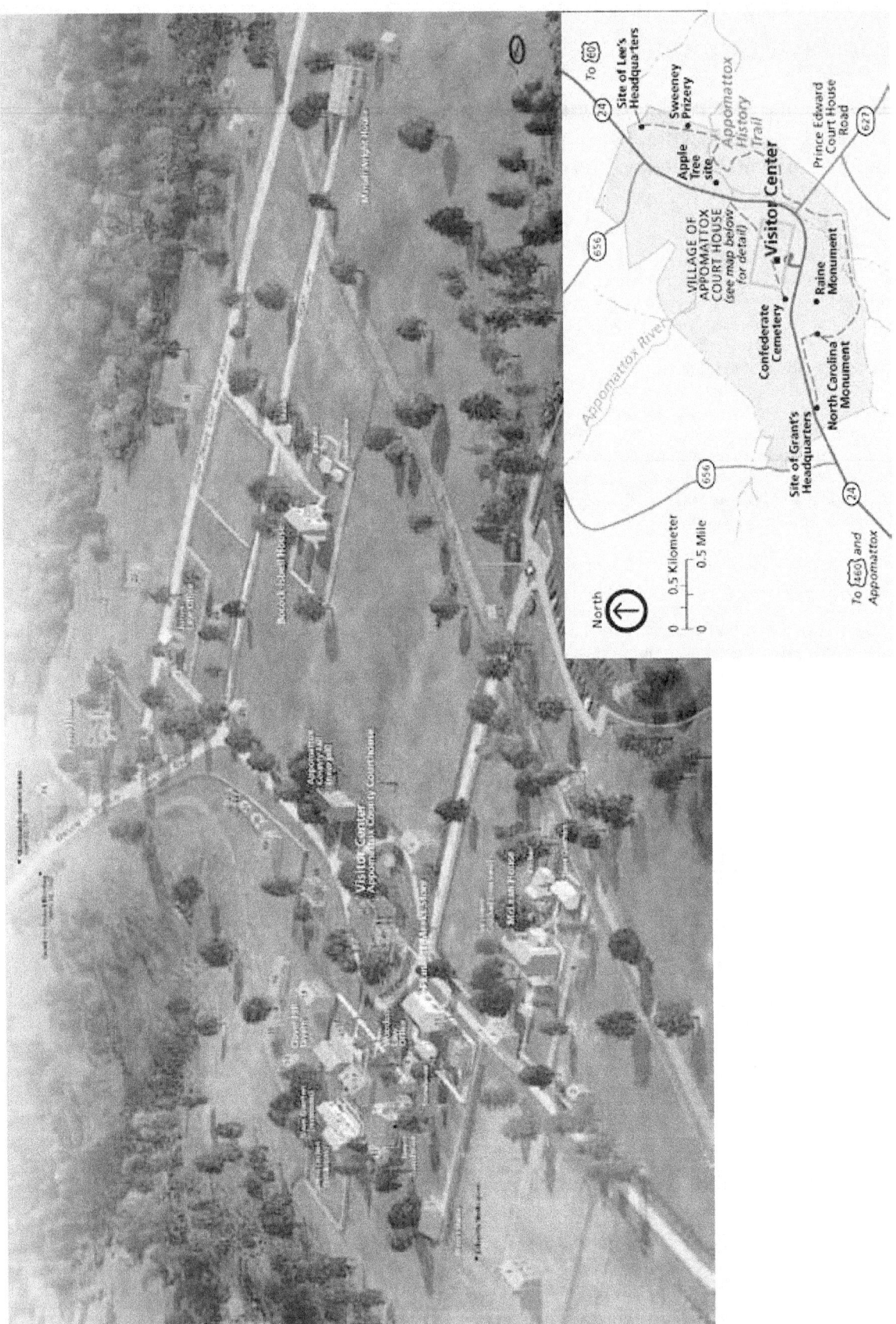

Figure 1. Maps of Appomattox Court House National Historical Park. Large map shows main visitor area of the park, historic structures, and illustrates the gentle rolling hills and forested character of the park. Smaller map shows boundary of park and location of major features, including the Appomattox River and its tributaries. Gray box shows location and extent of visitor area map. NPS Graphics.

Introduction

The following section briefly describes the National Park Service Geologic Resources Inventory and the regional geologic setting of Appomattox Court House National Historical Park.

Purpose of the Geologic Resources Inventory

The Geologic Resources Inventory (GRI) is one of 12 inventories funded under the NPS Natural Resource Challenge designed to enhance baseline information available to park managers. The program carries out the geologic component of the inventory effort from the development of digital geologic maps to providing park staff with a geologic report tailored to a park's specific geologic resource issues. The Geologic Resources Division of the Natural Resource Program Center administers this program. The GRI team relies heavily on partnerships with the U.S. Geological Survey, Colorado State University, state surveys, and others in developing GRI products.

The goal of the GRI is to increase understanding of the geologic processes at work in parks and provide sound geologic information for use in park decision making. Sound park stewardship relies on understanding natural resources and their role in the ecosystem. Geology is the foundation of park ecosystems. The compilation and use of natural resource information by park managers is called for in section 204 of the National Parks Omnibus Management Act of 1998 and in NPS-75, Natural Resources Inventory and Monitoring Guideline.

To realize this goal, the GRI team is systematically working towards providing each of the identified 270 natural area parks with a geologic scoping meeting, a digital geologic map, and a geologic report. These products support the stewardship of park resources and are designed for non-geoscientists. During scoping meetings the GRI team brings together park staff and geologic experts to review available geologic maps and discuss specific geologic issues, features, and processes.

The GRI mapping team converts the geologic maps identified for park use at the scoping meeting into digital geologic data in accordance with their Geographic Information Systems (GIS) Data Model. These digital data sets bring an interactive dimension to traditional paper maps by providing geologic data for use in park GIS and facilitating the incorporation of geologic considerations into a wide range of resource management applications. The newest maps come complete with interactive help files. As a companion to the digital geologic maps, the GRI team prepares a park-specific geologic report that aids in use of the maps and provides park managers with an overview of park geology and geologic resource management issues.

For additional information regarding the content of this report and up to date GRI contact information please refer to the Geologic Resources Inventory Web site (http://www.nature.nps.gov/geology/inventory/).

Park Setting

Geographic and Cultural Setting

Established by an Act of Congress in 1930, Appomattox Court House National Historical Park covers 705 ha (1,743 ac) in Appomattox County, Virginia. The park is situated some 16 km (10 mi) east of Lynchburg and 24 km (15 mi) west of Farmville. The park preserves the village of Appomattox Court House and surrounding farm, forest, and meadow lands (fig. 1). According to the enabling legislation, the mandated purpose of Appomattox Court House National Historical Park is to "commemorate the effective termination of the Civil War brought about by the surrender of the Confederate Army of Northern Virginia, under General Robert E. Lee to the Union Army under Lt. General Ulysses S. Grant at Appomattox Court House, Virginia on April 9, 1865 and for the further purpose of honoring those who engaged in this tremendous conflict."

The park preserves many cultural resources—including the site of the surrender meeting at McLean House, the historic village at Appomattox, and the site of the Battle of Appomattox Court House. Fought on April 9, 1865, the Battle of Appomattox Court House was a failed last-ditch effort by the Confederate army that led directly to the surrender.

Geologic Setting

Rolling hills and narrow eroded stream valleys characterize the landscape at Appomattox Court House National Historical Park. The elevation in the park ranges from 250 m (830 ft) above sea level at the highest hills to 194 m (638 ft) above sea level at the eroded Appomattox River valley. The Appomattox River and several tributaries, including Plain Run Branch (see inside front cover), traverse the park, which protects over 13 km (8.2 mi) of waterways. The Appomattox River is a major tributary to the James River. The two rivers meet in Hopewell, Virginia.

The park lies within the western Piedmont physiographic province. In the area of Appomattox, Virginia, the eastern United States is divided into the following five physiographic provinces, with associated local subprovinces (from east to west): the Atlantic Coastal Plain, the Piedmont, the Blue Ridge, the Valley and Ridge, and the Appalachian Plateaus provinces. Following is a general east-to-west description of the five physiographic provinces of this area (see fig. 2). The information is relevant to understanding the geologic history and current landscape setting of Appomattox Court House National Historical Park.

Atlantic Coastal Plain Province

The Atlantic Coastal Plain province is primarily flat terrain with elevations ranging from sea level to about 100 m (300 ft) in Virginia. Sediments eroding from the Appalachian Highland areas to the west formed the wedge-shaped sequence of soft sediments that were deposited intermittently during periods of higher sea level over the past 100 million years. These sediments are now more than 2,400 m (8,000 ft) thick at the Atlantic coast, and are reworked by fluctuating sea levels and the continual erosive action of waves along the coastline. Large streams and rivers in the Coastal Plain province—including the James, York, Rappahannock, and Potomac—continue to transport sediment and extend the coastal plain eastward. Beyond the province to the east, the submerged Continental Shelf province extends for another 121 km (75 mi).

Piedmont Province

The eastward-sloping Piedmont is located between the Blue Ridge province along the eastern edge of the Appalachian Mountains and the Atlantic Coastal Plain province to the east. The "Fall Line" (also known as the "Fall Zone") marks a transitional zone where the softer, less consolidated sedimentary rock of the Atlantic Coastal Plain province to the east intersects the harder, more resilient metamorphic rock to the west, forming an area of ridges, waterfalls, and rapids. Examples of the transition are present in the Potomac Gorge of the Chesapeake and Ohio Canal National Historical Park (Maryland, District of Columbia, and West Virginia). The Piedmont physiographic province encompasses the Fall Line westward to the Blue Ridge Mountains. The Piedmont is composed of crystalline igneous and metamorphic rocks such as schists, phyllites, slates, gneisses, and gabbros. This province formed through a combination of folds, faults, uplifts, accretions, and subsequent erosion. The resulting landscape of gently rolling hills, starting at an elevation of 60 m (200 ft), becomes gradually steeper toward the western edge where it reaches an elevation of 300 m (1,000 ft) above sea level. In the Appomattox area, the Piedmont is subdivided into three subprovinces, herein referred to as (from east to west): the Outer Piedmont subprovince; Mesozoic lowlands subprovince; and western Piedmont (also called Foothills subprovince), containing numerous distinct belts. Geologic provinces locally include the Chopawamsic Volcanic Belt, Carolina Slate Belt, and Goochland Terrane (fig. 3). Geologists frequently group these entities together into the Outer Piedmont subprovince. Several Mesozoic-age extensional basins (also called the Mesozoic lowlands subprovince) cut through the Piedmont locally, including through the Farmville basin east of Appomattox Court House (figs. 2 and 3).

Blue Ridge Province

The Blue Ridge province is located along the eastern edge of the Appalachian Mountains. The highest elevations in the Appalachian Mountain system occur within the Black Mountain subrange of the Appalachian Mountains northeast of Asheville, North Carolina. Proterozoic and Paleozoic (see fig. 8) igneous, sedimentary, and metamorphic rocks were uplifted during several orogenic events (i.e., Taconic, Acadian, and Alleghanian) to form the steep rugged terrain. Resistant Cambrian quartzite forms most of the high ridges, whereas Proterozoic metamorphic rocks underlie the valleys. The elongated belt of the Blue Ridge stretches from Georgia to Pennsylvania. Eroding streams have caused narrowing in the northern section of the Blue Ridge Mountains into a thin band of steep ridges that rise to heights of approximately 1,200 m (3,900 ft). These streams link the adjacent Valley and Range and Piedmont provinces.

Valley and Ridge province

Long parallel ridges separated by valleys (100 to 200 m [330 to 660 ft] deep) characterize the landscape of the Valley and Ridge physiographic province. The landforms are strongly representative of the lithology and structure of the deformed bedrock: valleys formed in easily eroded shale and carbonate formations among resistant sandstone ridges. The province contains strongly folded and faulted sedimentary rocks in western Virginia. The eastern portion is part of the Great Valley; this section is rolling lowland formed on folded carbonate rocks and shale.

Appalachian Plateaus Province

Compared to the more eastern physiographic provinces, the Appalachian Plateaus province is relatively undeformed. Instead of highly folded and inclined strata like the Valley and Ridge province, the rock layers of this province are nearly flat. A steep scarp, known as the Allegheny Front, bounds the plateau on the east. This escarpment rises abruptly 300 to 900 m (1,000 to 3,000 ft). Maximum elevations at this front are generally greater than that of the ridges in the Valley and Ridge province. Deep ravines carved into the horizontal sedimentary rock layers characterize the topography of this province. Geologic units are typically repeated sequences of shale, coal, limestone, and sandstone. Erosion of these units created a rugged, jumbled topographic surface.

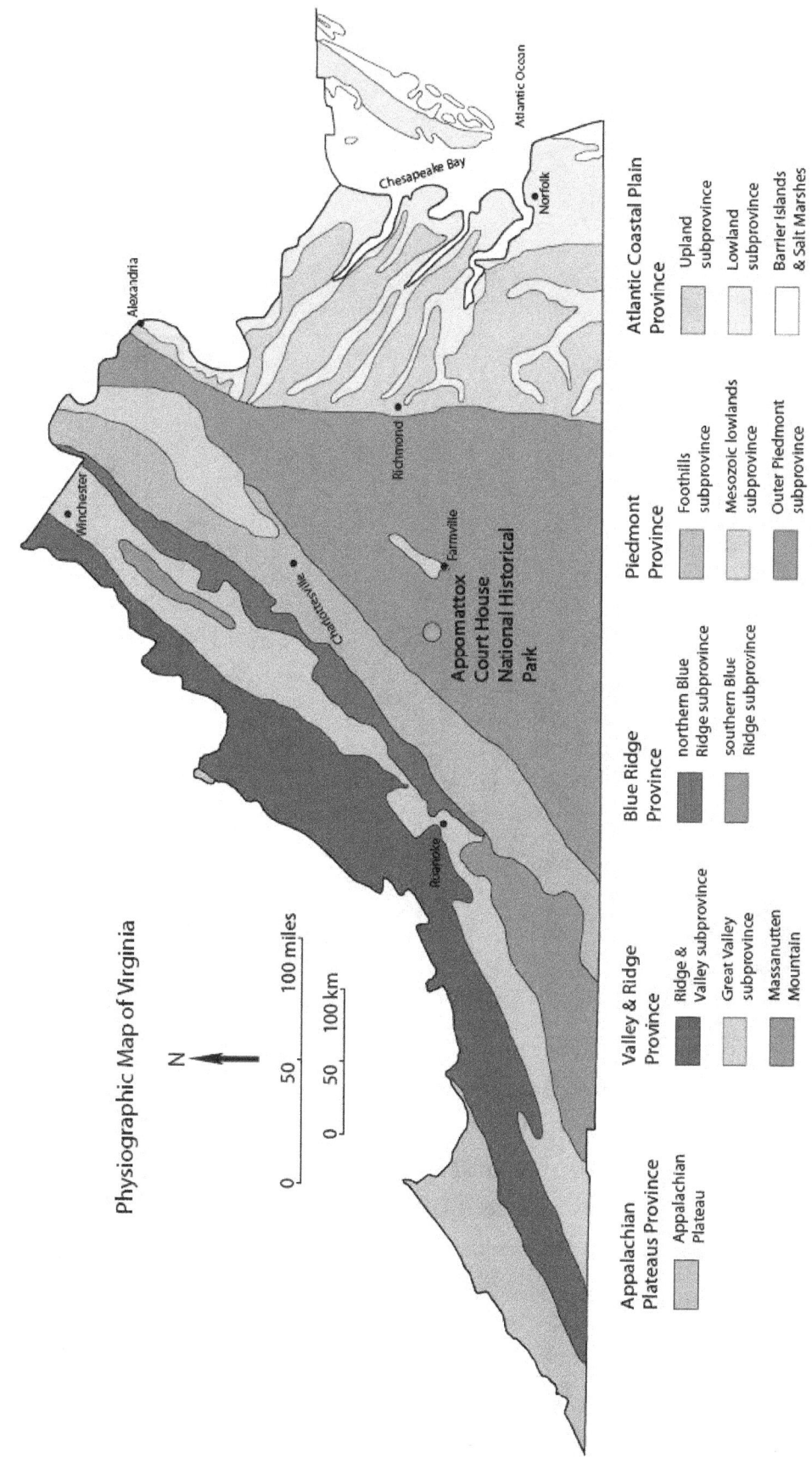

Figure 2. Location of Appomattox Court House National Historical Park relative to the physiographic provinces and subprovinces of Virginia. Graphic adapted from Bailey (1999).

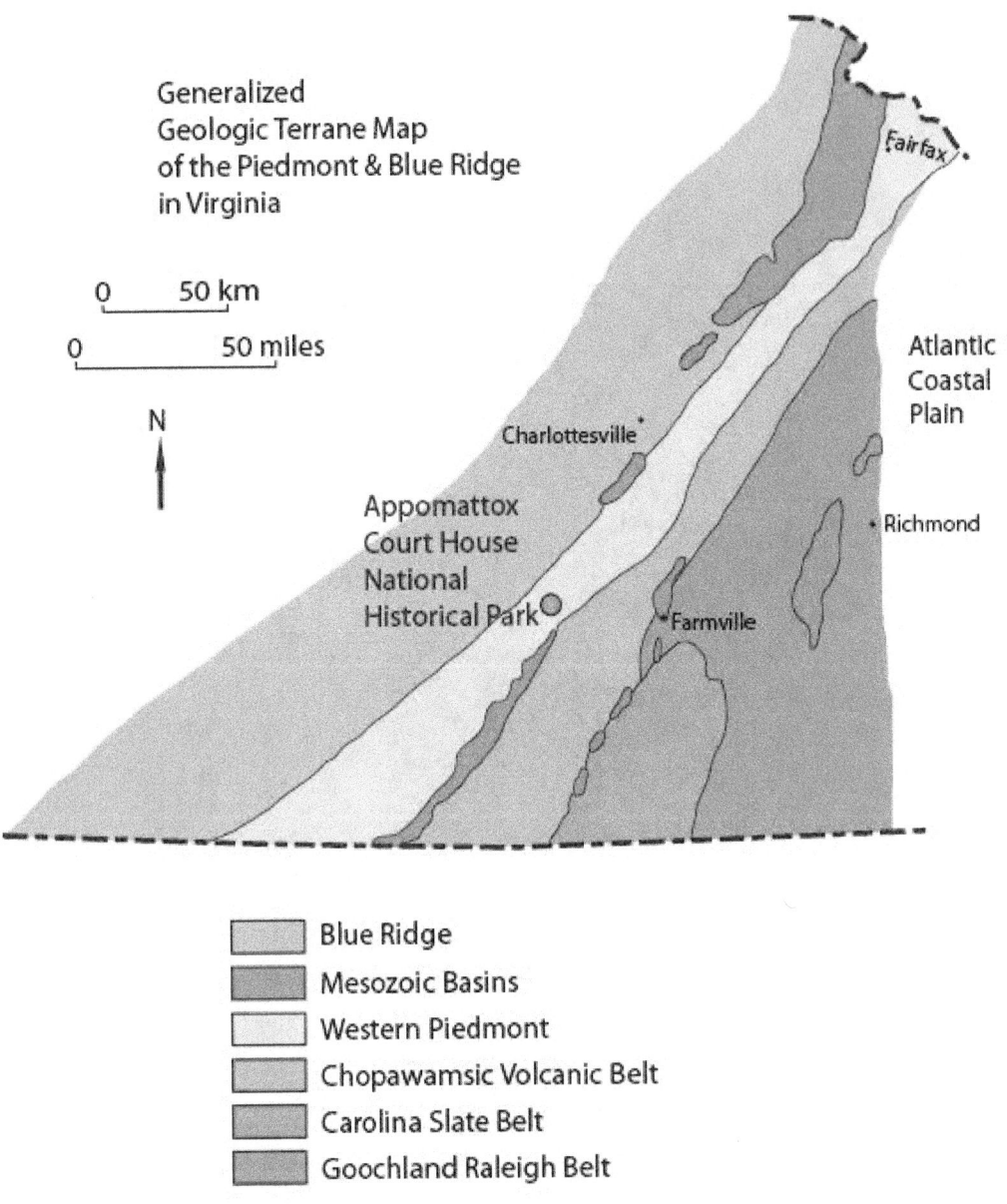

Generalized
Geologic Terrane Map
of the Piedmont & Blue Ridge
in Virginia

0   50 km

0   50 miles

N

Charlottesville

Appomattox
Court House
National
Historical Park

Farmville

Fairfax

Atlantic
Coastal
Plain

Richmond

Blue Ridge
Mesozoic Basins
Western Piedmont
Chopawamsic Volcanic Belt
Carolina Slate Belt
Goochland Raleigh Belt

Figure 3. Appomattox Court House National Historical Park relative to the geologic terranes of the Piedmont physiographic province of Virginia. Graphic adapted from Bailey (1999).

Geologic Issues

The Geologic Resources Division held a Geologic Resources Inventory scoping session for Appomattox Court House National Historical Park on April 21, 2005, to discuss geologic resources, address the status of geologic mapping, and assess resource management issues and needs. The following section synthesizes the scoping results, in particular those issues that may require attention from resource managers.

This section discusses natural resources management at Appomattox Court House National Historical Park. The issues are presented in order of importance; the most critical topics are discussed first, followed by potential research projects and other matters of scientific interest. Contact the Geologic Resources Division for assistance.

Surface Water Issues and Channel Morphology

Several waterways cross the landscape within and surrounding Appomattox Court House National Historical Park (fig. 1). Within the boundaries of the park are 13 km (8.2 mi) of streams, including the Appomattox River, North Branch, Plain Run Branch (see inside front cover), and Rocky Run. The largest waterway, the Appomattox River, is a major contributory to the James River and the Chesapeake Bay watershed. A portion of the headwaters of the Appomattox River is contained within the park boundaries. The total drainage area for the Appomattox River is 3,481 sq km (1,344 sq mi), with the park containing an important stretch of this system (Bell et al. 1996). Therefore, surface water quality within the park—and, by extension, in the surrounding communities—is very important to park resource managers.

The Clean Water Act Section 303 and Environmental Protection Agency regulations (40 CFR section 130.7) mandate a public list of all waters that do not support existing or designated beneficial uses, such as recreation and aquatic life support. Sources of pollution that frequently degrade water quality in the Appomattox Court House National Historical Park area include, but are not limited to, industrial and municipal point discharges, atmospheric input, and agricultural use such as fertilizers, pesticides, and animal waste (National Park Service 2007). Good water quality in the park is imperative for both the persistence of natural aquatic communities and the consumption and recreational use of water by park visitors.

Currently, the streams within the park boundaries meet state and federal Clean Water Act water quality standards, as reported by the Water Resources Division of the National Park Service (National Park Service 2007). However, increasing use and surrounding development threaten this status by ground compaction and increases in impervious surfaces such as parking lots and roadways. Impervious surfaces decrease the natural absorption of precipitation, funneling it as sheet flow directly into waterways.

In an ongoing effort to estimate loads of nutrients and suspended solids contributed by regional tributaries into Chesapeake Bay, the U.S. Geological Survey conducts regular water quality sampling. From 1993 through 1995, sampling along the Appomattox River near its confluence with the James River (east of Appomattox Court House National Historical Park) showed overall increases in total nitrogen, phosphorous, and total suspended solids in the Appomattox River. The causes of these trends remain unclear (Bell et al. 1996; Belval and Campbell 1996). Because much of the drainage basin is downstream from the park, diligence in maintaining good water quality, as well as determining the external sources of contamination, is vital to preserving a healthy aquatic system.

Flooding and channel erosion naturally occur along most of the streams and rivers within the park. Periodic flooding of the Appomattox River is the result of sudden large inputs of precipitation. Historically, regional rivers experienced major floods during all seasons of the year and especially during hurricane season. Major floods occur on average several times each decade (U. S. Army Corps of Engineers 1975). One the primary goals of the park is to present the historical context of the area, but channel erosion may alter the Civil War era features at Appomattox Court House. Alterations to park vegetation along exposed slopes may exacerbate channel morphology changes and stream bank erosion. For example, clearing trees (and their shoreline stabilizing roots) for historical restoration can lead to increased erosion, thereby increasing sediment load in nearby streams (GRI scoping meeting notes 2005).

Increased sediment load affects aquatic and riparian habitat in the Appomattox River and tributaries (M. Carter, written communication, December 2008). Sediments in the Appomattox River naturally contain heavy minerals derived from the local bedrock such as tourmaline, epidote, amphibole-pyroxene, staurolite, rutile, magnetite-ilmenite, muscovite, kyanite, zircon, garnet, sillimanite, biotite, chlorite, and titanite (Maccubbin 1952). As these minerals naturally break down, they may release heavy metals into the surrounding sediment and water. Sediment load typically increases during sudden heavy flow events; water turbidity also increases, and can cause the release of adsorbed contaminants and/or heavy metals contained in the sediment. Park resource management should be aware of stream sediment composition as it relates to aquatic and riparian habitat conditions.

Understanding baseline conditions of the aquatic ecosystem is vital to monitoring its future health. A 1976 study in Chesterfield County along the Appomattox River east of the park determined the taxonomy and ecology of algae as indicators of domestic sewage pollution, progress of wastewater treatment efforts, toxicity of industrial wastes, sources of surface water, and natural purification in streams (Woodson and Afzal 1976). The following data were collected as part of the study: 1) water pH to determine extent of buffering; 2) dissolved oxygen content to determine degree of aeration; 3) calcium, nitrogen, and phosphorous values; and 4) algal species distributions and characteristics. This type of study contributes to the understanding of the aquatic environment and, by extension, the natural buffering effects of the underlying geology and riverine sediments.

The Appomattox River and other local streams continuously change position as part of natural meandering river flow. These shoreline changes may threaten existing park facilities and distort the historical context of the landscape. Seasonal storm events, including microbursts and thunderstorms, have sent torrents of rain in localized areas across central Virginia. Such storms are responsible for debris flows along slopes of the Blue Ridge Mountains, and can cause slumping and landsliding on even moderate slopes within Appomattox Court House National Historical Park (GRI, scoping meeting notes, 2005).

Increased flow and flooding and sediment loading threaten several riparian wetlands within the park. Although these wetlands are small in scale at Appomattox Court House, they serve as indicators of overall ecosystem health and merit research and periodical monitoring.

Inventory, Monitoring, and Research Recommendations for Surface Water Issues and Channel Morphology

- Monitor erosion rates and shoreline changes along the Appomattox River, and compare to previous conditions using aerial photographs where available.
- Monitor the amount of runoff in the park, focusing on areas of recent development, to determine if direct runoff is increasing.
- Perform aquatic ecosystem (algal) surveys to determine the effect of increased erosion, sediment load, and sheet flow on aquatic ecosystems at the park.
- Inventory and monitor water and soil quality in wetlands to establish a basis for comparison of future conditions. Aerial photographs may provide data related to changes in wetland distribution through time.
- Investigate revegetation of vulnerable reaches of park streams to prevent excess erosion and sediment loading.
- Inventory runoff and flood-susceptible areas using paleoflood data and hydrologic regime modeling.
- Use shallow (25-cm; 10-inch) and deeper core data to monitor rates of sediment accumulation and erosion in the river, local streams, and springs.

Seismicity and Mass Wasting

Although the central Virginia area is not commonly associated with seismic activity, it does experience regular earthquakes, which are possibly due to crustal relaxation in which the Earth's crust adjusts to tectonic forces along deep-seated faults crossing the area (GRI, scoping notes, 2005; M. Carter, written communication, December 2008). In 2003, a 4.0- to 4.8-magnitude earthquake occurred in the area (GRI, scoping meeting notes, 2005). The ground shaking associated with earthquakes may trigger landslides, damage historic buildings and other man-made structures, and produce groundwater and surface water disturbances. The probability of a destructive seismic event at the park is low, but resource management should be aware of earthquake potential.

As mentioned above, seismicity can trigger mass wasting on even moderate slopes. Mass wasting is the downhill movement of soil and rock fragments induced by gravity. The subdued, rolling hill topography in the area appears stable; however, this area can experience large amounts of precipitation and hurricane events (GRI, scoping notes, 2005), and the likelihood of landslides and slumps increases with heavy precipitation as well as by undercutting of slopes by streams, roads, trails, and other infrastructure. Erosion near streams causes increased sediment load and gullying, and can threaten or destroy trails, bridges, and other features of interest. During natural river meandering, higher erosion rates along the outer portions of bends in streambeds (where stream velocity is higher) causes the bank to retreat, undercutting the bank and leading to washout. Remedial structures such as cribbing, log frame deflectors, jack dams, stone riprap, and log dams armor the bank, deflect the flow, and help slow erosion (Means 1995). Where appropriate, these techniques could protect historic and cultural resources at Appomattox Court House National Historical Park.

Shrink-and-swell clays occur in the substrate in the park area. These clays expand when saturated and shrink when dry. Clays are locally derived from weathered amphibolite-containing rocks, such as those present beneath the park. In extreme cases, shrink-and-swell clays can damage roads, buildings, and other structures. Similarly, these slippery clays can affect visitor trails, causing them to be unsafe in wet conditions.

Inventory, Monitoring, and Research Recommendations for Seismicity and Mass Wasting

- Cooperate with local universities and agencies such as the U.S. Geological Survey to monitor seismic activity in the area.
- Determine the vulnerability index of slopes for failure during intense seasonal storm events.
- Monitor erosion rates along slopes within the park by establishing key sites for routine profile measurements and photographs to document rates of erosion or deposition; also measure changes after major storm events.
- Using factors such as different sediment deposit compositions, slope aspects, and vulnerability indices,

perform a comprehensive study of the erosion and weathering processes active at the park.

- Use a topographic map to determine steepness, a geologic map to determine rock type, and rainfall information to determine the relative potential (risk) for landslides to occur on slopes throughout the park.
- Map locations of shrink-and-swell clay units to avoid for future development. Any exposed layers of these clay units should be avoided or covered for trails.

Anthropogenic Impacts

Since the early 1700s, human settlement practices throughout the area created an impacted landscape that persists today at Appomattox Court House National Historical Park. Removals of trees, soil, and rocks, as well as grazed pastures and homestead sites, are evident across the landscape. Many of these older features are part of the preserved cultural context in the park. However, like much of central Virginia, the area surrounding the park is becoming increasingly populated. As urban and suburban development continues, conservation of any existing natural forest-meadow community types becomes a critical concern. The park exists as a protected area, ecologically connected to the surrounding areas. External impacts may negatively affect ecosystems within the park.

Human impacts continue as water lines, gas lines, power lines, radio towers, industrial complexes (along the southern boundary of the park), roads, and housing developments increase on the landscape. Cattle grazing on leased lands within and adjacent to the park threaten surface water and groundwater quality. Overused, overgrazed, and deforested riparian corridors increase stream bank erosion and suspended sediment in local waterways. Additionally, trails, visitor use areas, imported (invasive) species, acid rain, and air and water pollution damage the landscape. Resource management of these impacts is an ongoing process at the park.

Inventory, Monitoring, and Research Recommendations for Anthropogenic Impacts

- Using geologic unit descriptions and a detailed geologic map (1:6,000-scale preferably), determine if any buffering of acidic precipitation is occurring. Focus on areas containing calcareous rocks. Relate this information to the water quality for the park.
- Inventory and consider monitoring soils and bedrock to determine the extent of chemical alterations and pH changes through time.
- Seek out cooperative arrangements with surrounding landowners and local developers and industry to minimize impacts near park areas and to promote environmentally sound development and land use practices.
- Quantitatively determine effects of cattle grazing by performing repeated stream channel surveys along grazed areas and downstream locations.

Mine Features

Throughout the history of the Appomattox Court House National Historical Park area, humans have sought to exploit its geologic resources. Appomattox lent its name to a coal field (now a potential source of methane) within the Richmond basin along the Appomattox River near Petersburg, east of the park (Tuomey 1842; Topuz et al. 1982). Within the park boundaries, a historical depression, now flooded as a vernal pool, has enigmatic origins (fig. 4; GRI, scoping notes, 2005). The surrounding geologic units provided attractive building stone from local quarries. Nearby stone quarries provided building stones for Civilian Conservation Corps road-building during the Great Depression of the 1930s. In addition, small-scale abandoned sand and gravel pits dot the landscape around the park. If these features are important historic relicts, they could be an interpretive program target. Local mining is also discussed in the "Geologic Features and Processes" section. The Virginia Division of Mineral Resources has an updated inventory and database of mine features such as small open pit mines (GRI, scoping meeting notes, 2005). This database does not indicate the presence of any mines within Appomattox Court House National Historical Park (M. Carter, written communication, December 2008).

Inventory, Monitoring, and Research Recommendations for Mine Features

- Inventory any mine-related features at the park, consulting aerial photographs and historic records. Determine if further monitoring of each site is necessary.
- Further investigation into the origin of the vernal pool may yield additional information regarding any historical significance and connections to the hydrogeologic system.

Hydrogeologic Modeling

Because water is such a vital natural resource, resource managers should understand how water is moving through the hydrogeologic system into, under, and from Appomattox Court House National Historical Park. The bedrock beneath the park is deeply weathered, deformed, fractured, and faulted (Nelson et al. 1999). Fractures, joints, and cracks provide quick conduits for water flow and strongly affect the hydrologic system. Within 10 m (33 ft) of the surface, the bedrock can be highly weathered and much more permeable (Virginia Division of Mineral Resources 2003). Knowledge of the hydrogeologic system is critical to understanding the impacts of human-induced contaminants on the ecosystem, as well as predicting groundwater table changes. The groundwater flow system and all interactions between groundwater flow, surficial water flow, and overall water quality at the park should be quantitatively determined. Little data exist on the nature of the hydrogeologic system at the park.

Nearby cattle grazing threatens the groundwater quality in the area with point sources of animal waste as well as increased erosion and sediment load in surface streams. Several wells throughout the park could be used for monitoring of groundwater quality (GRI, scoping meeting notes, 2005). Dye-tracer studies using these wells could greatly illuminate the hydrogeologic system underlying the park to determine how quickly and in what direction water is moving through the system. In 1997, the U.S. Geological Survey performed groundwater modeling to determine the hydrogeologic framework, analyze groundwater flow, and understand regional flow in the fall zone near Richmond, east and northeast of Appomattox Court House (McFarland 1997a, 1997b). The modeling methodology involved the installation of a regular network of observation wells to supplement existing wells. The network consisted of three well transects from which hydrogeologic sections were modeled. Hydrogeologic data included aquifer material sample descriptions (geologic logs), well depths, gamma logs, water levels, aquifer horizontal hydraulic conductivities (using single-well "slug" tests), age of groundwater, pH, temperature, specific conductance, and dissolved oxygen. With these data, groundwater discharge at various selected seepage sites, as well as vertical hydraulic conductivity, could be determined (McFarland 1997a, 1997b). Hydrogeologic modeling (including field studies and local-scale flow-model analyses) requires the establishment of baseline conditions and repeated monitoring (McFarland 1997a, 1997b). This type of study facilitates understanding the hydrogeologic framework and its relationship with the surficial ecosystem.

Inventory, Monitoring, and Research Recommendations for Hydrogeologic Modeling

- Inventory and monitor groundwater levels throughout the park using existing wells, and install new wells as necessary. Include and relate to well-located natural springs and seeps wherever possible.
- Test for and monitor organics (from agricultural and cattle waste), phosphate, and volatile hydrocarbon levels in groundwater at the park. Focus on areas near visitor and park management facilities and at boundaries near industrial sites and housing developments.
- Work with the National Park Service Water Resources Division to create hydrogeologic models for the park to better manage the groundwater resource and predict the system's response to contamination.

Future Facility Planning

As stated earlier, geologic features and processes such as shrink-and-swell clays, erosion, flooding, and seismicity are present and active in the Appomattox Court House area. Understanding these factors is important in deciding the locations and parameters of future facilities at the park. The development of visitor use sites, including trails and picnic areas, are now being considered by park managers. High concentrations of shrink-and-swell clays could undermine any structure and create an unsafe trail base when wet. Knowing the locations of springs and understanding the hydro-geologic system at the park helps resource managers avoid problems without interrupting groundwater flow. Similarly, the siting of waste treatment facilities requires accurate hydrogeologic models in place to avoid unnecessary groundwater contamination.

Inventory, Monitoring, and Research Recommendations for Future Facility Planning

- Consult local geologists familiar with the hydrogeologic system, topographic expression, and land use history of the area when planning future facilities at the park.
- Locate areas of shrink-and-swell clays, springs, high groundwater flow, relatively steep slope, unstable substrate, and undercut slopes to avoid for future facilities development.

Miscellaneous Action Items

- Evaluate the feasibility of supporting detailed (1:6,000-scale) geologic mapping of surficial and bedrock geology within park boundaries.
- Develop an interpretive program illuminating the balance between cultural and historical context and natural processes at the park.
- Update the park website and other interpretive products to include connections between geologic content and other scientific (biological) and cultural disciplines.

Figure 4. The "Tibbs vernal pool" in Appomattox Court House National Historical Park. This unique feature, of unknown origin, is underwater in early spring (A; photo taken in March 2003). By the summer months, the pool is dry (B). Note the highwater marks on the trees. NPS photos courtesy Brian Eick (Appomattox Court House NHP).

Figure 5. McLean House, the surrender site. Photo by Landry C. Thornberry.

Figure 6. Appomattox County Courthouse. Photo by Landry C. Thornberry.

Geologic Features and Processes

This section describes the most prominent and distinctive geologic features and processes in Appomattox Court House National Historical Park.

Geologic Structures

Appomattox Court House National Historical Park sits within the Western Piedmont subprovince, which is separated from the Central Virginia Volcanic-Plutonic Belt (also called the Chopawamsic Volcanic Belt) by several reactivated thrust faults and Mesozoic extensional basins, including the Farmville and Danville basins to the southeast, the Brookneal fault (shear) zone, and the Shores mélange zone. The Brookneal shear zone, named for the town of Brookneal, Virginia approximately 13 km (8 mi) southwest of Appomattox, is a dextral fault zone of late Paleozoic age characterized by wide swaths of shear bands and deformed metamorphic rocks such as mylonite and cataclasite (Nelson et al. 1999). On the western edge of Appomattox County, the Western Piedmont abuts the Proterozoic-age rocks of the Blue Ridge physiographic province. The rocks in this area tend to occur in elongated parallel belts trending northeast-southwest, dipping steeply to the southeast, and reflecting the accretion of terranes throughout the development of eastern North America (Nelson et al. 1999; Hackley et al. 2007).

The rocks of the area have a complex deformational and metamorphic history with at least four distinct periods of folding (Marr and Sweet 1980). Large-scale regional folds include the Whispering Creek anticline and the Long Island syncline, both located east of Appomattox (Marr 1981). Recent mapping south of Appomattox Court House National Historical Park revealed a northern extension of the Carolina Slate Belt along a simple upright northeast-trending isoclinal syncline called the Dryburg syncline (fig. 7). It is part of the Virgilina synclinorium complex that stretches south into North Carolina and contains greenschist facies, metamorphosed volcanic, volcaniclastic, and sedimentary rocks (Hackley et al. 2007). Folding and metamorphism of this structure likely occurred during the Alleghanian Orogeny, as described in the "Geologic History" section.

Appomattox Court House National Historical Park sits atop bedrock that includes the Fork Mountain Formation; the bedrock contains porphyroblastic aluminosilicate mica schist, garnet-bearing biotite gneiss, and amphibolite, as well as a band of greenstone and amphibole gneiss, with an interlayered belt of felsic metatuff, mica schist, and gneiss (Virginia Division of Mineral Resources 2003). Permeating through the regional rock units are small-scale faults, fractures, mylonite zones, foliation, and other deformation features. Foliation is commonly wavy, defined by thin alternating dark and light mineral bands (Nelson et al. 1999). Small-scale joints and fractures are ubiquitous in the rocks beneath the park. Most deformation fabrics follow the northeast-southwest regional structural trend.

This may be due to overprinting of Alleghanian Orogeny deformation textures during the later Mesozoic extensional events.

Parallel thrust faults, trending northeast-southwest, cut through the Appomattox County area (Bailey 2000). Many of these faults reactivated as normal faults during Mesozoic extension and may still accommodate crustal stresses. Prominent among them is the Bowens Creek fault, which is a Paleozoic high-strain zone. This sheared zone transitions farther north of Appomattox into the Mountain Run zone and trends northeast just east of Charlottesville (Bailey 2000). The Mountain Run shear zone separates the Western Piedmont from the Blue Ridge physiographic province. East of Appomattox, several thrusts, reactivated as reverse faults, dip to the east and parallel the regional structural trend (Dischinger 1987; Budke 1992).

Geology, Biology, and History Connections

Geologic resources played key roles in the settlement and development of the Appomattox Court House National Historical Park area. Rocks in eastern Appomattox County contain massive sulfide deposits of early mining interest (Marr 1981; Good 1981). These sulfides (lead, zinc, iron, manganese, and copper) are part of the gold-pyrite belt of northern and central Virginia Piedmont and part of the larger Appalachian-Caledonide North American metallogenic province, which extends from Alabama northeast to Newfoundland. These sulfide deposits originated on the sea floor, formed by hot brines or fumaroles and preserved in lavas, volcanic tuffs, and marine sediments adjacent to island arcs formed in late Proterozoic, Cambrian, or Ordovician time (Good 1981). Kyanite (an alumino-silicate mineral formed at high pressure) was also mined by power shovel at Baker Mountain, east of Madisonville, which is 8 km (5 mi) southeast of Appomattox (Nelson et al. 1999).

Highly weathered rocks with deep weathering profiles and sparse bedrock exposure underlie most of the Appomattox area (Nelson et al. 1999). The geology, in direct correlation with the soil types and distributions present at the park, also controls natural biologic patterns in trees and other plants as well as agricultural success patterns. The geologic units at Appomattox Court House National Historical Park include mafic metavolcanics and amphibolites. These calcium-rich rocks weather to create fertile soils that attract specific native plants such as the American red cedar and redbuds, which prefer calcium-rich substrates. Other areas within Appomattox County have variable regoliths with abundant clay and kaolinite (Thomas et al. 1989a). These substrates are relatively impermeable and not ideal agricultural settings. Variability in soils reflects the variety of geologic units underlying the area, including metagraywackes and mafic igneous intrusions (Thomas et al. 1989a). Abundant local clay contributed to industry, with Appomattox County once containing the largest manufacturer of ceramic smoking pipes (Appomattox Historical Society 2007).

The first inhabitants to this area were the Appomattox Indians. Land clearing, agriculture (tobacco farms), and early settlement focused on the landscape surrounding Appomattox Court House as early as the 18th century as settlers received land grants from the Colonial Governor (Appomattox Historical Society 2007). Appomattox County formed in 1845 with the first courthouse at Clover Hill Village, later renamed Appomattox Courthouse.

The rolling hills and gentle landscape and topography at Appomattox Court House National Historical Park are defined by the local geology and hydrology. This setting historically dictated the placement of towns, strategies and encampment of troops, escape routes, river crossings, and railroads, as well as the development of outlying areas. In many Civil War maneuvers, an advantage was held by those familiar with the terrain; the ability to utilize the natural features of the area and manipulate the focal points, gaps, ravines, cuts, hills, and ridges could determine the outcome of a battle. In addition to influencing battles, the landscape and topography also affected how troops and supplies were transported during the Civil War. Because Appomattox Court House is famous for the site of Lee's surrender, geologic controls on the landscape and Civil War story can be overlooked during interpretation.

One of the major goals of the park is to present the historical context of the area; this includes preserving and restoring historic structures that remain and the landscape around them. So-called cultural landscapes provide the physical environment associated with historical events and reveal aspects of the country's development through their form, features, and use, in addition to illustrating the relationships among park cultural and natural resources. The National Park Service Cultural Landscape Inventory contains information on the location, historical development, and current management of cultural landscapes, including condition.

The cultural landscapes for Appomattox Court House National Historical Park are not officially identified or listed on the Cultural Landscapes Inventory. Because the primary feature within the park is the restored 19th century village of Appomattox Court House, the park contains a least one if not several cultural landscapes (National Park Service 2007). The more than 55 historic structures, including the McLean House (surrender site) and Appomattox County Courthouse, within the park are an integral part of this cultural landscape (figs. 1, 5, and 6).

Maintaining this Civil War landscape often means resisting natural geologic changes, and this presents several management challenges. Geologic processes such as chemical weathering, seismicity, and slope erosion are constantly changing the landscape at the park. Runoff erodes sediments from any open areas and carries them down streams, ravines, and gullies. Erosion naturally diminishes higher areas such as ridges and hills, foundations, and earthworks. Erosion degrades bridge foundations and cuts streams back into restoration areas; subsequent deposition of added sediments fills in the lower areas such as trenches, railroad cuts, and stream ravines, distorting the historical context of the landscape. Alterations to park vegetation along exposed slopes lead to changes in the hydro-geologic regime at the park. For example, clearing trees and their stabilizing roots for historical restoration can lead to increased erosion, thereby increasing sediment load in nearby streams and undermining slopes.

Interpreters make the landscape come alive for visitors and give the scenery a deeper meaning. Because the geology of Appomattox Court House is intimately connected with its biology and human history, it would be appropriate to emphasize the geologic features, processes, and influences on landscape development at the park. Interpretive displays and programs initiated by cooperative efforts between local geologists, biologists, and interpretive staff would contribute to the visitors' experience, deepening a feeling of connection between park visitors and the landscape. One possible product of such a cooperative effort could be a geologic general interest map for visitors that contains simple explanatory text focusing on the geologic influences on the American Civil War battles and troop movements throughout the area.

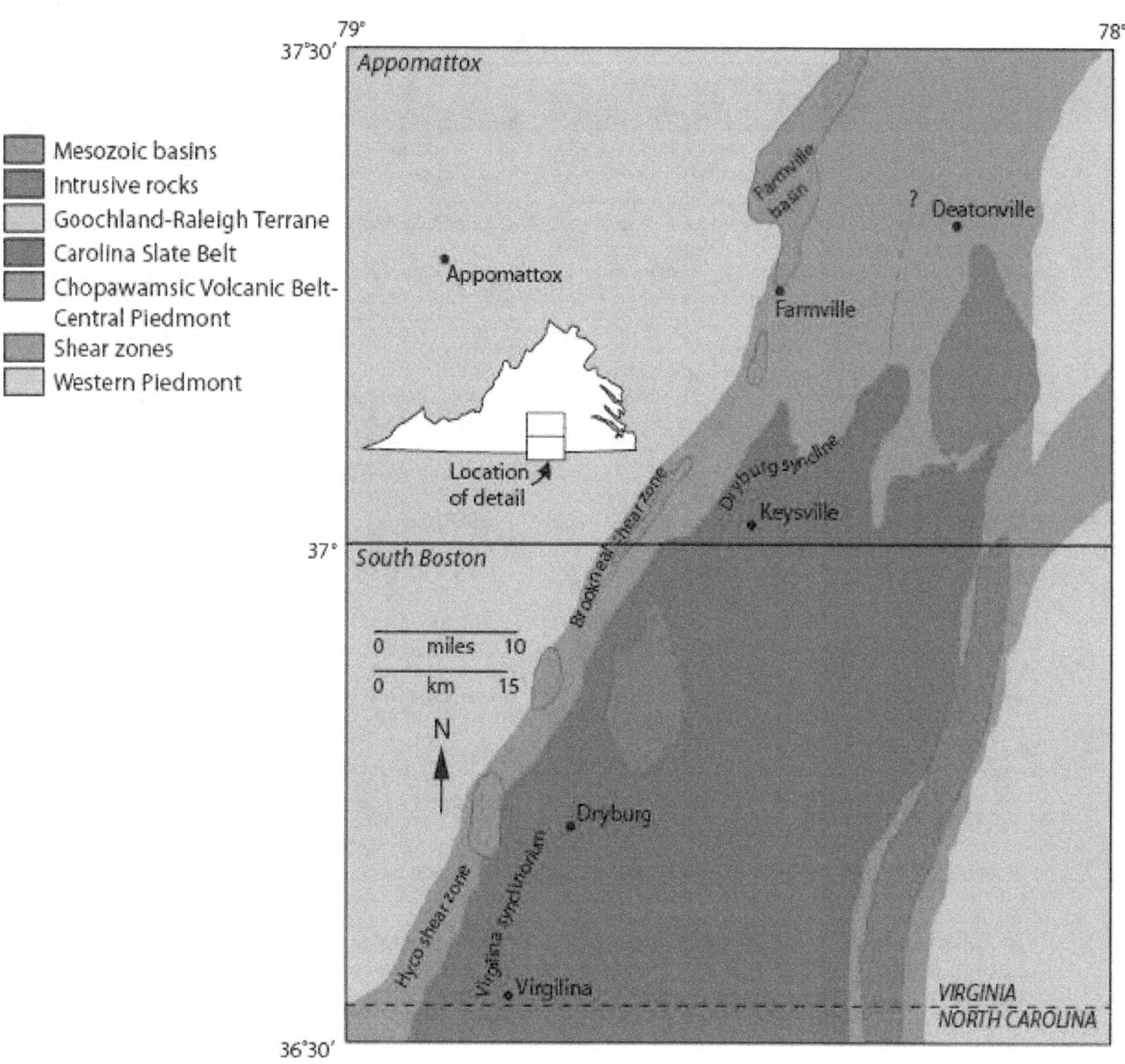

Legend:
- Mesozoic basins
- Intrusive rocks
- Goochland-Raleigh Terrane
- Carolina Slate Belt
- Chopawamsic Volcanic Belt-Central Piedmont
- Shear zones
- Western Piedmont

Figure 7. Generalized geologic map of Appomattox and South Boston 30- x 60-minute quadrangles in south central Virginia. Figure shows the boundaries of major tectonic elements, the location of the Dryburg syncline (red line), and regional towns. Graphic adapted from Hackley et al. (2007) by Trista L. Thornberry-Ehrlich (Colorado State University).

Map Unit Properties

This section identifies characteristics of map units that appear on the Geologic Resources Inventory digital geologic map of Appomattox Court House National Historical Park. The accompanying table is highly generalized and is provided for background purposes only. Ground-disturbing activities should not be permitted or denied on the basis of information in this table.

Geologic maps facilitate an understanding of the Earth, its processes, and the geologic history responsible for its formation. Hence, the geologic map for Appomattox Court House National Historical Park informed the "Geologic History," "Geologic Features and Processes," and "Geologic Issues" sections of this report. Geologic maps are essentially two-dimensional representations of complex three-dimensional relationships. The various colors on geologic maps represent rocks and unconsolidated deposits. Bold lines that cross and separate the color patterns mark structures such as faults and folds. Point symbols indicate features such as dipping strata, sample localities, mines, wells, and cave openings.

Incorporation of geologic data into a geographic information system (GIS) increases the utility of geologic maps and clarifies spatial relationships to other natural resources and anthropogenic features. Geologic maps are indicators of water resources because they show which rock units are potential aquifers and are useful for finding seeps and springs. Geologic maps do not show soil types and are not soil maps, but they do show parent material, a key factor in soil formation. Furthermore, resource managers have used geologic maps to make correlations between geology and biology; for instance, geologic maps have served as tools for locating threatened and endangered plant species, which may prefer a particular rock unit.

Although geologic maps do not show where future earthquakes will occur, the presence of a fault indicates past movement and possible future seismic activity. Geologic maps will not show where the next landslide, rockfall, or volcanic eruption will occur, but mapped deposits show areas that have been susceptible to such geologic hazards. Geologic maps do not show archaeological or cultural resources, but past peoples may have inhabited or been influenced by various geomorphic features that are shown on geologic maps. For example, alluvial terraces may preserve artifacts, and inhabited alcoves may occur at the contact between two rock units.

The features and properties of the geologic units in the following table correspond to the accompanying digital geologic data. Map units are listed from youngest to oldest. Please refer to the geologic time scale (fig. 8) for the age associated with each time period. This table highlights characteristics of map units such as susceptibility to hazards; the occurrence of fossils, cultural resources, mineral resources, and caves; and the suitability as habitat or for recreational use.

GRI digital geologic maps reproduce essential elements of the source maps including the unit descriptions, legend, map notes and graphics, and report. The following are source data for the GRI digital geologic map for Appomattox Court House National Historical Park:

Virginia Division of Mineral Resources. 2003. Digital representation of the 1993 geologic map of Virginia. Publication 174 [CD-ROM; 2003, December 31]. Richmond, VA: Commonwealth of Virginia Department of Mines, Minerals, and Energy.
Adapted from:
Virginia Division of Mineral Resources. 1993. Geologic map of Virginia and Expanded Explanation. Scale 1:500,000. Richmond, VA: Commonwealth of Virginia Department of Mines, Minerals, and Energy.

The GRI team implements a geology-GIS data model that standardizes map deliverables. This data model dictates GIS data structure including data layer architecture, feature attribution, and data relationships within ESRI ArcGIS software, increasing the overall quality and utility of the data. GRI digital geologic map products include data in ESRI personal geodatabase, shapefile, and coverage GIS formats, layer files with feature symbology, FGDC metadata, a Windows HelpFile that contains all of the ancillary map information and graphics, and an ESRI ArcMap map document file that easily displays the map. GRI digital geologic data are included on the attached CD and are available through the NPS Data Store (http://science.nature.nps.gov/nrdata/).

Map Unit Discussion

Appomattox Court House National Historical Park is in the western portion of the Piedmont physiographic province. This portion of the Piedmont contains mixed metamorphosed and deformed distal marine sediments and mid-oceanic rift igneous rock once deposited in the Iapetus Ocean basin (see "Geologic History" section). The oldest rocks in the greater area include Late Proterozoic to Cambrian age metagraywacke, quartzose schist, granite, banded marble, amphibolite, and gneiss (Virginia Division of Mineral Resources 2003).

The oldest rocks in the Appomattox Court House National Historical Park crop out along the deepest eroded reaches of the Appomattox River. Bedrock units within the park include the Fork Mountain Formation, which contains porphyroblastic aluminosilicate mica schist, garnetiferous biotite gneiss, and amphibolite. Other units appearing within the park boundaries include a band of greenstone and amphibole gneiss, with an interlayered belt of felsic metatuff, mica schist, and gneiss (Virginia Division of Mineral Resources 2003).

Rocks mapped in the surrounding areas (up to ≈ 13 km [8 mi] away) include Cambrian-age foliated felsite, Ordovician-age biotite granite gneiss, and garnet-biotite schist. Mappable units of Mesozoic mylonite and diabase record pervasive deformation and igneous intrusion in the Appomattox area (Virginia Division of Mineral Resources 2003).

The youngest units (not indicated on the bedrock digital geologic map) at the park include thick alluvium deposits of sand, gravel, silt, and clays; marsh and swamp deposits along larger rivers; terrace deposits atop higher areas; and artificial fill from construction of roads, dams, bridges, landfills, and highways.

The following pages present a tabular view of the stratigraphic column and an itemized list of features for each map unit. Map units are listed from youngest to oldest; please refer to the geologic time scale (fig. 8) for the age associated with each time period. This table includes properties of each map unit such as map symbol, name, description, resistance to erosion, suitability for development, hazards, potential paleontological resources, cultural and mineral resources, potential karst issues, potential for recreational use, and global significance.

Map Unit Properties Table

Colored rows indicate units mapped within Appomattox Court House National Historical Park.

Age	Unit Name (Symbol)	Features and Description	Erosion Resistance	Suitability for Development	Hazards	Cultural Resources	Karst	Mineral Occurrence	Habitat	Geologic Significance
LOWER JURASSIC	Diabase (Jd)	Unit typically appears as irregular masses, but may also occur as discrete dikes and sills. In Appomattox area, unit is linear. Diabase is an intrusive igneous rock with fine- to coarsely-crystalline, subaphanitic or porphyritic with aphanitic margins. Crystalline textures include dark-gray mosaic of long thin plagioclase crystals and clinopyroxene. Some masses contain olivine or bronzite, whereas others are granophyric.	Very high if unweathered.	Suitable for most development unless high, fractured, or weathered (friable), in which case there is a potential for shrink and swell clays. Unweathered rocks of this unit are nearly impossible to modify or move without blasting and heavy equipment.	Rockfall potential for this unit if exposed on slope.	Unit often underlies ridges that factored into troop movements throughout the Civil War.	None.	Plagioclase, clinopyroxene. Unit may be used for decorative stone.	Unit may support localized plant species changes.	Unit is time-correlative with Mesozoic extension events across the eastern United States.
MESOZOIC	Mylonite (My)	Deformed metamorphic rock. Textures include protomylonite, mylonite, ultramylonite, and cataclasite. Myriad lithologies present, depending on the nature of the parent rock, and degree and nature of deformation. Most outcrops contain belts of mylonite and cataclastic rock anastomosing around lenses of less-deformed or undeformed country rock. Typically, mylonitic and cataclastic rocks are gradational into less deformed or undeformed adjacent rocks, and contacts are approximate or arbitrary. Most belts of mylonite represent fault zones with multiple deformational events throughout the Paleozoic and Mesozoic locally.	Moderate to low depending on degree of deformation and weathering.	Units are heavily deformed and foliated, containing small lenses of many rock types. This variability may render the units unstable for heavy development. The high mica content of this unit gives it the potential for severe erosion and sedimentation control may be required.	Deformation bands between large blocks may cause them to be susceptible to rockfall and mass wasting on slopes.	None documented.	None.	None documented.	Deformed nature of unit creates localized permeability changes that may affect vegetation patterns.	Unit contains deformation fabrics from Paleozoic contractional events superimposed on Late Precambrian extensional and locally contractional fabrics. Mesozoic extensional reactivation overprints earlier fabrics along local faults.
ORDOVICIAN	Porphyroblastic garnet-biotite schist (Oas)	Mica schist contains 1- to 2-mm (0.04-0.08 in) garnet porphyroblasts in an anastomosing, greenish-black biotite-rich, schistose matrix. Many exposures show complex microstructures recording many phases of deformation. Locally, unit contains quartz-rich muscovite schist and thin interbeds of calcareous mica schist and marble.	Moderate.	Avoid highly schistose layers as well as areas of intense preferential compositional weathering (along foliation). High mica contents render the unit susceptible to severe erosion and causes increased sediment load in regional streams.	Lenses of resistant rocks in schistose layers may be susceptible to mass wasting. Severe erosion and sedimentation potential exists if unit is exposed due to high mica content. Potential for shrink and swell clay issues.	Garnets provide abrasive material and may have been objects of early trade.	Some dissolution is possible in local marbles.	Biotite, garnet, muscovite, quartz, plagioclase, magnetite, kyanite, and calcite. Some marble present.	Unit weathers to produce iron- and calcium-rich soils.	Unit has elongate positive magnetic and radiometric anomalies as an enigmatic geophysical signature.
ORDOVICIAN	Lineated biotite granite gneiss (Lgn)	Gneiss ranges from light-colored to -equal amounts of light and dark minerals, with medium- to coarse-grained textures. Lineation fabrics are strong and some areas are locally porphyritic. Unit comprises more than one intrusive body and interlayers with Cambrian-age rocks associated with volcanic processes.	Moderate to high.	Units are suitable for most development unless local blocks and/or plagioclase groundmass are altered and weathered to saprolite, rendering the unit friable. Unit may be deeply weathered and unstable for deep excavated cuts.	Avoid units exposed and undercut on slopes due to potential spalling and mass wasting hazards.	None documented.	None.	Quartz, potassium feldspar, plagioclase, biotite, muscovite, garnet.	None documented.	Western portions of this unit coincide with a strong positive radiometric anomaly.
CAMBRIAN	Felsic gneiss, schist, and metasiltstone (Cfs)	*(text too faded to read reliably)*	Moderate.	*(text too faded to read reliably)*	*(text too faded to read reliably)*	Unit exposed in the park and outcrops may have Civil War significance.	None.	Quartz, plagioclase, muscovite, biotite, epidote, magnetite.	Unit has resistant, coarse-crystalline regolith with low permeability.	Unit records Cambrian sedimentary and volcanic processes.
CAMBRIAN	Greenstone or amphibolite gneiss (Cma)	Greenstone is dark to dusky-green, with schistose textures including actinolite-chlorite schist mineralogies with segregations of the chlorite to epidote and quartz. Greenstone is interbedded with hornblende-plagioclase gneiss, and subordinate amounts of detrital metatuff, quartz-muscovite schist, and fine-grained salt-and-pepper biotite-muscovite gneiss.	Moderate.	Intersection of bedding and flow cleavages in greenstones as well as heavily altered zones may be points of weakness in unit. Dusty clay, slick surfaces, and swell potential may shrink and swell clay potential; avoid development of buildings and septic fields on this unit.	Lenses of resistant amphibolites in schistose layers may be susceptible to block and rockfall. Abundant shrink and swell clay potential exists for this unit.	Unit exposed in the park and outcrops may have Civil War significance.	None.	Actinolite, chlorite, epidote, quartz, hornblende, plagioclase, muscovite, biotite.	Unit weathers to produce calcium- and magnesium-rich, smectitic clay soils, well-suited for growing tobacco.	Unit records subaqueous basaltic extrusion during the Cambrian.

Age	Unit Name (Symbol)	Features and Description	Erosion Resistance	Suitability for Development	Hazards	Cultural Resources	Karst	Mineral Occurrence	Habitat	Geologic Significance
CAMBRIAN	Melrose granite (Cm)	Granite is light greenish gray with pink bands present in irregular masses. Textures range from medium to coarse grained. Compositions vary between quartz monzonite and quartz diorite. Near the Brookneal shear zone, deformation textures in the granite grade from protomylonite to mylonite to ultramylonite.	High to moderate where deformed.	Suitable for most development unless highly weathered and/or fractured. Avoid weathered areas for basements and foundations due to potential radon problem.	May be susceptible to mass wasting if exposed on slope.	Coarse-grained specimens may have attracted early trade interest.	None.	Quartz, plagioclase, potassium feldspar, biotite, muscovite, chlorite, epidote, titanite, garnet, magnetite-ilmenite, calcite, zircon.	Deformed nature of unit creates localized permeability changes that may affect vegetation patterns.	Unit has a positive-radiometric/negative-magnetic geophysical signature and a radiometric (Uranium-Lead) age date of 515 million years.
CAMBRIAN	Foliated felsite (Cfv)	Light-colored igneous rock. Foliated unit ranges in composition from rhyolite to dacite. In outcrop, unit appears light gray to white with medium-grained textures.	Moderately high.	Avoid weathered areas of rhyolite composition for basements and foundations due to potential radon problem. Unit may weather to saprolite and would be severely susceptible to erosion due to high mica content.	May be susceptible to mass wasting if exposed on slopes. Severe erosion and sedimentation potential exists if unit is exposed, due to high mica content.	Cryptocrystalline metarhyolite was a prized tool material for American Indians.	None.	Quartz, perthitic microcline, muscovite, biotite. Large beta-form quartz phenocrysts.	Unit weathers to produce potassium-rich soils.	Unit records Cambrian-age volcanism.
CAMBRIAN	Amphibolite, hornblende-biotite gneiss and schist (Cmv)	In outcrop, unit appears black to moderate olive brown. Lineations and foliations record pervasive deformation with textures ranging from medium to coarse grained. Stringers of quartz and epidote are relatively common.	Moderately high.	Suitable for most development unless highly weathered and/or fractured. Development of buildings and other infrastructure may be compromised due to the presence of abundant shrink and swell clays associated with weathered portions of this unit.	Preferential weathering of less resistant layers may increase likelihood of spalling when unit is exposed on a slope. Abundant shrink and swell clays associated with this unit can cause widespread instability.	Coarse-grained specimens may have attracted early trade interest.	None.	Hornblende, tremolite-actinolite, oligoclase, biotite, epidote, garnet.	Unit weathers to produce "sweet" calcium-rich soils.	Unit include widely recognized Blackwater Creek Gneiss and Catawba Creek amphibolite member of Hyco Formation, hornblende gneiss, and dominantly mafic-composition units.
LATE PROTEROZOIC-CAMBRIAN	Banded marble (PCZac)	Unit appears light and dark gray with medium-grained textures and fine laminations. Other interlayered rock types include calcareous gneiss and schist. Locally thick to thin beds of marble alternate with graphitic phyllite and mica schist, with compositions ranging from impure marble to calcareous metagraywacke.	Moderate.	Unit is suitable for most development unless highly heterogeneous layers are present, or beds are heavily fractured. Avoid areas with high carbonate contents for wastewater treatment or septic systems.	Unit is susceptible to mass wasting when exposed on slopes and undercut by local rivers. Pyrite and other sulfide minerals could produce acid drainage if disturbed.	None documented.	Karst dissolution is possible in calcareous units.	Calcite, quartz, biotite, muscovite, plagioclase, pyrite, magnetite-ilmenite.	Unit weathers to produce calcium-rich soils.	Unit records terrane accretion along continental margins and includes the Arch Marble and Archer Creek Formation.
LATE PROTEROZOIC-CAMBRIAN	Feldspathic metagraywacke (PCZmy)	Myriad rock types include medium to light gray laminated quartz- and feldspar-rich to calcareous gneiss with thin mica schist partings; white and gray fine- to coarse-grained generally laminated marble; gray to greenish gray fine-grained graphitic mica schist and quartzite; light gray medium- to fine-grained mica schist; massive quartzite and micaceous blue quartz granule metasandstone; and dark greenish black actinolite schist.	Moderate.	Avoid areas of intense preferential compositional weathering (along foliation). Suitable for most development unless highly weathered and/or fractured, especially along talc-rich layers and boundaries. Actinolite schists may weather to produce shrink and swell clays.	May be susceptible to mass wasting if exposed on slope, talc and weathered areas may increase likelihood of sliding. Graphitic schists may contain high sulfide assemblages that could produce acid drainage if disturbed.	None documented.	Some dissolution of laminated marble is possible.	Quartz, potassium feldspar, plagioclase, biotite, muscovite, calcite, epidote, titanite, magnetite-ilmenite, graphite, chlorite, albite, garnet, kyanite, tremolite, talc, actinolite, dolomite.	None documented.	Units record multiple deformation events, including uplift along the Bowens Creek fault; they were previously mapped as the Evington Group and are continuous with the Lynchburg Group.
LATE PROTEROZOIC-CAMBRIAN	Fork Mountain Formation (PCZfm)	Unit contains aluminosilicate-mica schist interlayered with other metamorphic rocks such as garnet-rich biotite gneiss, calcsilicate granofels, amphibolite, rare white marble, and calc-quartzite lenses. Schist is light to medium gray, with textures ranging from fine to medium grained. Units underwent multiple episodes of deformation and were metamorphosed with locally emergent porphyroblasts. Metamorphic and deformation textures include schistosity, multiple crenulation cleavages, and partly retrograded porphyroblasts of garnet and aluminosilicate. Some brittle deformation is recorded in polymictic breccias.	Mostly moderate depending on degree of alteration and deformation.	Unit has highly heterogeneous lithology and is heavily deformed, rendering it rather weak for heavy development. Amphibolite layers may weather to produce shrink and swell clays that may pose problems for building foundations and other infrastructure.	High degree of deformation renders unit unstable on slopes; may be prone to mass wasting. Shrink and swell clays associated with weathering of amphibolites in this unit can cause widespread instability.	Garnets provide abrasive material and may have been objects of early trade. Unit is exposed in the park and outcrops may have Civil War significance.	Rare white marbles may be prone to dissolution.	Quartz, muscovite, biotite, garnet, staurolite, magnetite, ilmenite, rutile, paragonite, plagioclase, sillimanite, potassium feldspar, tourmaline, hornblende, epidote, kyanite, chlorite, chloritoid, andalusite, corundum, spinel.	Deformed nature of unit creates localized permeability changes that may affect vegetation patterns.	Unit has characteristic "curly maple" magnetic contour map patterns due to isolated concentrations of highly magnetic minerals. Many mineral assemblages record prograde and retrograde metamorphic events in the area.

Age	Unit Name (Symbol)	Features and Description	Erosion Resistance	Suitability for Development	Hazards	Cultural Resources	Karst	Mineral Occurrence	Habitat	Geologic Significance
LATE PROTEROZOIC – CAMBRIAN	Metagraywacke, quartzose schist, and mélange (CZpm)	Metagraywackes include quartz-rich chlorite or biotite schists, and contain very fine to coarse granules of blue quartz. Shearing transposes primary graded laminations into elongate lozenge shapes that give the rock a distinctive pinstriped appearance in weathered surfaces that are perpendicular to schistosity. Locally, a mylonitic fabric and late-stage chevron-shaped folds overprint the earlier schistosity, and rocks in this unit are progressively more deformed from east to west across the outcrop belt. The most deformed rocks contain mylonitic mica schist with quartz-rich sausage-shaped segments. Rock fragments contain dacite tuff, gabbro, and monocrystalline quartz with zircon and biotite inclusions. These blocks range from 5 cm to 3 m (2 in to 10 ft) across in outcrop.	Moderate depending on degree of deformation.	Schistose, mylonitic, and highly foliated units should be avoided for heavy development due to inherent weakness. Heavily deformed and/or weathered areas are also unsuitable and unstable in deep excavated cuts.	May be susceptible to mass wasting if exposed on slope. Weathered material (saprolite) is susceptible to severe erosion and sedimentation problems if exposed.	Blue quartz may have been used for early trade.	None.	Quartz, albite, epidote, chlorite, muscovite, magnetite, chloritoid, calcite, biotite, staurolite, plagioclase, perthite, tourmaline, titanite.	Unit weathers to produce relatively impermeable acidic, orange, saprolitic regolith; quartz-rich schists weather to a light yellowish-gray, sandy, mica-rich saprolite.	Locally polydeformed quartz-rich mica schists are lithologically indistinguishable from schists mapped as Fork Mountain Formation (CZfm) in structural blocks that occur to the west, and may be temporally correlative.

Geologic History

This section describes the rocks and unconsolidated deposits that appear on the digital geologic map of Appomattox Court House National Historical Park, the environment in which those units were deposited, and the timing of geologic events that created the present landscape.

Proterozoic Eon

The recorded geologic history of the Appalachian Mountains begins in the Proterozoic Eon (figs. 8, 9, and 10A). In the mid-Proterozoic, during the mountain building event called the Grenville Orogeny, a supercontinent formed that incorporated most of the continental crust in existence at that time, including the crust of today's North America and Africa. The sedimentation, deformation, plutonism (the intrusion of igneous rocks), and volcanism are manifested in the metamorphic gneisses in the core of the modern Blue Ridge Mountains west of Appomattox Court House National Historical Park (Harris et al. 1997). These rocks were deposited over a period of 100 million years and are more than a billion years old, making them among the oldest rocks known from this region. They form a basement upon which all other rocks of the Appalachians were deposited (Southworth et al. 2001).

The late Proterozoic (fig. 10B), roughly 600 million years ago, brought a rifting (pulling apart) tectonic setting to the area. The supercontinent broke up and a sea basin formed that eventually became the Iapetus Ocean. This basin collected many of the sediments that would later form the Appalachian Mountains and Piedmont Plateau. Mixed sediments (now metamorphosed) such as the Fork Mountain Formation, metagraywacke, mélange, banded marble, and quartzose schist in the Appomattox area, accumulated during this time (Virginia Division of Mineral Resources 2003).

In addition, in this tensional environment, flood basalts and other igneous rocks such as diabase and rhyolite accumulated on the North American continent. These igneous rocks were intruded through cracks in the granitic gneisses of the Blue Ridge core and extruded onto the land surface during the breakup of the continental land mass (Southworth et al. 2001). The weathered and altered remains of the early flood basalts are preserved in the Catoctin Formation, located to the northwest of Appomattox Court House National Historical Park (Mixon et al. 2000).

Late Proterozoic Eon–Early Paleozoic Era

Associated with the shallow marine setting along the eastern continental margin of the Iapetus basin were large deposits of sands, silts, and muds in near-shore, deltaic, barrier island, and tidal flat areas (fig. 10C) (Schwab 1970; Kauffman and Frey 1979; Simpson 1991). As the Iapetus Ocean basin expanded, mid-ocean rift volcanism (similar to the modern Atlantic spreading ridge) mixed basaltic lavas with marine sediments. In the Appomattox area, the metamorphic remnants of these

mixed sedimentary rocks are the foliated felsites, amphibole gneiss, mica schist, and greenstone (Virginia Division of Mineral Resources 2003). In addition, huge masses of carbonate rocks represent a grand platform, thickening to the east, that persisted during the Cambrian and Ordovician periods (545 to 480 million years ago) (fig. 11A).

Somewhat later, in pulses occurring approximately 540, 470, and 360 million years ago, amphibolite, granodiorite and pegmatite, and lamprophyre, respectively, intruded the sedimentary rocks of the basin and nearshore areas. Several episodes of mountain building and continental collision responsible for the Appalachian Mountains contributed to the heat and pressure that deformed and metamorphosed the entire pile of sediments, intrusives, and basalts into schists, gneisses, marbles, slates, and migmatites (Southworth et al. 2001). Many of these rocks are preserved in the greater Appomattox Court House National Historical Park area.

From Early Cambrian through Early Ordovician time, orogenic activity along the eastern margin of the North American continent began again. This involved the closing of the Iapetus Ocean, subduction of oceanic crust, creation of volcanic arcs, and uplift of continental crust (fig. 11B) (Means 1995). The Taconic Orogeny (≈ 440 to 420 million years ago in the central Appalachians) was a volcanic arc–continent convergence. Oceanic crust, basin sediments, and the volcanic arc from the Iapetus Ocean basin were thrust onto the eastern edge of the North American continent. The initial metamorphism of the widespread Catoctin Formation and other basalts into greenstones, metabasalts, and metarhyolites, as well as the basin sediments (graywackes) into quartzites, schists, gneisses, and phyllites, occurred during this orogenic event. Rocks on the eastern edge of Appomattox County include a 5-to 8-km-wide (3- to 5-mi-wide) mélange assemblage of metagraywacke and metamorphosed mafic, ultramafic, and volcanic rocks within magnetite-bearing schists that were likely shoved against the North American continent at this time (Brown and Pavlides 1981). Pervasive regional metamorphism culminated in the Appomattox Court House National Historical Park area during this orogenic event (Nelson et al. 1999).

In response to the overriding plate thrusting westward onto the continental margin of North America, the crust bowed downward, creating a deep basin that filled with mud and sand eroded from the highlands to the east (Harris et al. 1997). This so-called Appalachian basin was centered on what is now West Virginia (fig. 11C).

Sandstones, shales, siltstones, quartzites, and limestones were continuously deposited in the shallow marine to deltaic environment of the Appalachian basin. This shallow marine to fluvial sedimentation continued for a period of about 200 million years during the Ordovician, Silurian, Devonian, Mississippian, Pennsylvanian, and Permian periods. This resulted in thick piles of sediments. The source of these sediments was from the highlands that were rising to the east during the Paleozoic orogenic events.

During the Late Ordovician, the oceanic sediments of the shrinking Iapetus Ocean were thrust westward onto other deepwater sediments of the western Piedmont along the Pleasant Grove fault. These rocks, now metamorphosed, currently underlie the Valley and Ridge province located west of the Blue Ridge (Fisher 1976).

Middle Paleozoic Era

The Acadian Orogeny (≈ 360 million years ago) continued the mountain building of the Taconic Orogeny as the African continent approached North America pushing ocean basin mixed sediments and volcanic rocks westward (Harris et al. 1997). Similar to the preceding Taconic Orogeny, the Acadian involved landmass collision, mountain building, and regional metamorphism (Means 1995). This event was focused farther north than south-central Virginia.

The Piedmont metasediments record the transition from non-orogenic, passive margin sedimentation to extensive, synorogenic (occurring during an orogeny) clastic sedimentation from the southeast during Ordovician time (Fisher 1976). In the Appomattox Court House National Historical Park area, these metasediments include schists, metagraywackes, phyllonites, gneisses, mélanges, and metasiltstones. Oceanic crust caught up in the orogenic events now exists in the Piedmont Plateau as peridotites, metagabbros, serpentinite, and pyroxenites, among other metamorphosed mafic rocks (Drake et al. 1994; Mixon et al. 2000).

Late Paleozoic Era

Following the Acadian orogenic event, the proto-Atlantic Iapetus Ocean was completely destroyed during the Late Paleozoic as the North American continent collided with the African continent. This formed the overall trend of the current Appalachian mountain belt, as well as a supercontinent named Pangaea. This mountain building episode is called the Alleghanian Orogeny (≈ 325 to 265 million years ago), and it was the last major orogeny of the Appalachian evolution (fig. 12A) (Means 1995). Coincident with continental collision was the deformation by folding and faulting that produced the large-scale Appalachian structures such as the Sugarloaf Mountain anticlinorium and the Frederick Valley synclinorium in the western Piedmont, the Blue Ridge-South Mountain anticlinorium, and the numerous folds of the Valley and Ridge province (Southworth et al. 2001). Faulting and deformation along the Brookneal shear zone southwest of Appomattox culminated during this event (Nelson et al. 1999).

During the Alleghanian Orogeny, rocks of the Great Valley (Shenandoah), Blue Ridge, and Piedmont provinces were transported as a massive block (Blue Ridge-Piedmont thrust sheet) westward onto younger rocks of the Valley and Ridge along the North Mountain fault. The amount of compression was very large. Estimates of the horizontal shortening range from 20 to 50 percent, which translates into 125 to 350 km (75 to 125 mi) of lateral translation (Harris et al. 1997).

Deformed rocks in the eastern Piedmont, east of Appomattox Court House National Historical Park, are composed of a number of terranes with different origins and geologic histories (Spears et al. 2004). East of the western Piedmont area is the Chopawamsic terrane. The Chopawamsic terrane is an Ordovician-age volcanic and plutonic island arc complex containing accompanying sedimentary rocks. All of these rocks have undergone regional metamorphism and deformation. Metamorphic conditions during the Alleghanian locally overprinted, but did not obliterate earlier Taconic Orogeny features (Nelson et al. 1999).

Located east of the Chopawamsic terrane, farther north of Appomattox Court House National Historical Park, is another exotic terrane accreted onto the North American Continent: the Goochland terrane. This body of rock is described as Middle to Late Proterozoic basement massif. It contains gneiss, amphibolite, granite, and anorthosite. The rocks in this terrane underwent multiple stages of deformation as well as granulite facies (high pressure and temperature) metamorphism during its emplacement, and subsequent squeezing throughout the Alleghanian Orogeny (Spears et al. 2004; Hackley et al. 2007). Separating these two terranes is a highly deformed band of rocks known as the Spotsylvania high-strain zone (Spears et al. 2004).

The Carolina Slate Belt contains slate, felsic crystal metatuff, biotite granodiorite, and hornblende diorite. These rocks were metamorphosed to greenschist facies (and higher grades northward) during the Alleghanian Orogeny (Horton et al. 1999; Hackley et al. 2007). The slate belt wedges between the Chopawamsic and Goochland terranes south of Appomattox Court House National Historical Park. The northern limit of this belt is unclear, but recent mapping extended it farther north along the Dryburg syncline (Hackley et al. 2007). The Chopawamsic and Goochland terranes, Carolina Slate Belt (located east of the park), and western Piedmont were also folded and faulted, and existing thrust faults were reactivated as both strike slip and thrust faults during the Alleghanian orogenic events (Southworth et al. 2001). Characteristic of the boundaries between these terranes are bands of strongly sheared rocks called mylonites and nearly melted migmatites (Horton et al. 1999). These rocks form under intense heat and pressure during deformation.

Wide bands of deformed mylonite occur in the Appomattox area. These mylonites have highly variable lithologies, and typically represent fault zones with multiple movement histories. Locally, these banded deformed rocks contain evidence of transpressional deformation that occurred during Late Paleozoic collisional tectonics. These features were then reactivated during later Mesozoic extensional faulting, as described below (Virginia Division of Mineral Resources 2003).

Millions of years of erosion now expose the metamorphosed core of the mountain range, but paleoelevations of the Alleghanian Appalachian Mountains are estimated at approximately 6,100 m (20,000 ft), analogous to the modern day Himalaya Range in Asia. For comparison, the highest peaks of Shenandoah National Park, west of Appomattox Court House National Historical Park are about 1,200 m (4,000 ft). Elevations within Appomattox Court House National Historical Park do not exceed 250 m (about 830 ft).

Mesozoic Era

Following the Alleghanian Orogeny, during the late Triassic, a period of rifting began as the deformed rocks of the joined continents began to break apart from about 230 to 200 million years ago (fig. 12B). The supercontinent Pangaea was segmented into roughly the continents that persist today. This episode of rifting or crustal fracturing initiated the formation of the current Atlantic Ocean, and caused many small-scale block-fault basins to develop with accompanying volcanism (Harris et al. 1997; Southworth et al. 2001). These Mesozoic basins are scattered around the park area and include the early Mesozoic Farmville and Taylorsville basins, present beneath a thick cover of younger deposits (Mixon et al. 2000).

The Triassic-age Newark Basin system is a large component of this rifting tectonic setting. This system of rift basins runs from New Jersey and Pennsylvania southwestward along a roughly linear trend toward North Carolina. Streams feeding large alluvial fans carried sediments and debris shed from the recently uplifted Blue Ridge and Piedmont provinces, and deposited them into fault-created troughs such as the nearby early Mesozoic Culpeper basin in the western Piedmont. These deposits were nonmarine shales, siltstones, and sandstones. Many of these rifted openings became lacustrine basins for thick deposits of siltstones and sandstones.

The large faults that formed the western boundaries of the basins provided an escarpment that was quickly covered with eroded debris. In addition to rifting, igneous rocks such as diabase intruded into the new strata as sub-horizontal sheets, or sills, and near-vertical dikes that extend beyond the basins into adjacent rocks. Diabase in this area tends to be rich in iron (Gottfried et al. 1991).

After these molten igneous rocks were emplaced during the Jurassic, approximately 200 million years ago, the region underwent a period of slow uplift and erosion. The uplift was in response to isostatic adjustments within the crust, which forced the continental crust upward and exposed it to further erosion (fig. 12C). Because it is harder than the surrounding sedimentary rocks, the igneous rocks (mostly diabase) in the park area were more resistant to erosion and now cap some of the higher ridges, hills, and slopes in the region.

Throughout the Mesozoic Era, thick deposits of unconsolidated gravel, sand, and silt eroded from the mountains. These were deposited at the base of the mountains as alluvial fans and spread eastward, covering metamorphic and igneous rocks and now form the Atlantic Coastal Plain (Duffy and Whittecar 1991; Whittecar and Duffy 2000; Southworth et al. 2001). The Cretaceous-age Potomac Formation, through which the Appomattox River cuts to the east, is a widespread example of a clastic wedge of shed sediment. The amount of material inferred from the grade of the now-exposed metamorphic rocks is immense. Many of the rocks exposed at the surface today must have been at least 20 km (≈ 10 mi) below the surface prior to regional uplift and erosion.

Since the regional uplift of the Appalachian Mountains and the subsequent breakup of Pangaea, the North American plate has continued to drift toward the west, creating an eastern passive margin. The isostatic adjustments that uplifted the continent after the Alleghanian Orogeny continued at a subdued rate throughout the Cenozoic period (Harris et al. 1997). These adjustments may be responsible for occasional seismic events felt across the region. The weathering and erosion continued throughout the Cenozoic Era with the early Potomac, Rappahannock, Appomattox, and James rivers carving their channels, stripping the Coastal Plain sediments, lowering the mountains, and depositing alluvial terraces along the rivers, forming the present landscape.

Cenozoic Era

The landscape and geomorphology of the greater Potomac, Rappahannock, and other large river valleys is the result of erosion and deposition from about the mid-Cenozoic period to the present, or at least the past 5 million years. The distribution of flood plain alluvium and ancient fluvial terraces blanketing the areas along the rivers and adjacent tributaries record the historical development of the drainage systems (Marr 1981). There is little to no evidence that the rivers migrated laterally across a broad, relatively flat region. It appears that the rivers have cut downward through very old resistant rocks, overprinting their early courses (Southworth et al. 2001). Regional concurrence of linear topographic and structural features, subsurface lineaments, and perched terraces suggest that recent deposition and landscape development have been tectonically controlled (Dischinger 1987).

Although not included on the 1:500,000-scale digital geologic map, many surficial deposits of alluvium cover the landscape of Appomattox Court House National Historical Park. The position, distribution, thickness, and elevation of alluvium and low-lying alluvial terraces, and the sediments deposited on them along the rivers, vary by province, age, and rock type. The elevations of terraces along the regional larger rivers show that the slope values of the ancient and modern river valleys are similar, which suggests that the terraces formed as the result of either global sea level drop or uplift (Zen 1997a, 1997b).

Although continental glacial ice sheets from the Pleistocene "ice ages" never reached the south-central Virginia area (the southern terminus was in northeastern Pennsylvania), the intermittent colder climates of the ice ages played a role in the formation of the landscape at Appomattox Court House National Historical Park. Freeze and thaw cycles at higher elevations in the Blue Ridge led to increased erosion of large boulders and rocks by ice wedging. Sea level fluctuations during ice ages throughout the Pleistocene caused the base level of many of the area's rivers to change. During lowstands (sea level drops), the rivers would erode their channels, exposing the deformed bedrock of the Piedmont Plateau. During oceanic highstands, the river basins flooded and deposition resulted in deposits of beach sediments far west of current shorelines.

Eon	Era	Period	Epoch	Ma	Life Forms	North American Events
Phanerozoic	Cenozoic	Quaternary	Holocene	0.01	Modern humans	Cascade volcanoes (W)
			Pleistocene		Extinction of large mammals and birds	Worldwide glaciation
		Neogene	Pliocene	2.6	Large carnivores	Sierra Nevada Mountains (W)
			Miocene	5.3	Whales and apes	Linking of North and South America
		Paleogene	Oligocene	23.0		Basin-and-Range extension (W)
			Eocene	33.9		
			Paleocene	55.8	Early primates	Laramide Orogeny ends (W)
				65.5		
	Mesozoic	Cretaceous			Mass extinction / Placental mammals / Early flowering plants	Laramide Orogeny (W) / Sevier Orogeny (W) / Nevadan Orogeny (W)
		Jurassic		145.5	First mammals	Elko Orogeny (W)
		Triassic		199.6	Mass extinction / Flying reptiles / First dinosaurs	Breakup of Pangaea begins / Sonoma Orogeny (W)
	Paleozoic	Permian		251	Mass extinction / Coal-forming forests diminish	Supercontinent Pangaea intact / Ouachita Orogeny (S) / Alleghanian (Appalachian) Orogeny (E)
		Pennsylvanian		299	Coal-forming swamps / Sharks abundant / Variety of insects / First amphibians	Ancestral Rocky Mountains (W)
		Mississippian		318.1	First reptiles	
		Devonian		359.2	Mass extinction / First forests (evergreens)	Antler Orogeny (W) / Acadian Orogeny (E-NE)
		Silurian		416	First land plants	
		Ordovician		443.7	Mass extinction / First primitive fish / Trilobite maximum / Rise of corals	Taconic Orogeny (E-NE)
		Cambrian		488.3	Early shelled organisms	Avalonian Orogeny (NE) / Extensive oceans cover most of North America
				542		
Proterozoic		Precambrian			First multicelled organisms	Formation of early supercontinent / Grenville Orogeny (E)
				2500	Jellyfish fossil (670 Ma)	First iron deposits / Abundant carbonate rocks
Archean					Early bacteria and algae	
				≈4000		Oldest known Earth rocks (≈3.96 billion years ago)
Hadean					Origin of life?	Oldest moon rocks (4–4.6 billion years ago)
				4600	Formation of the Earth	Formation of Earth's crust

Life Forms age groupings (vertical labels): Age of Mammals, Age of Dinosaurs, Age of Amphibians, Fishes, Marine Invertebrates

Figure 8. Geologic time scale; adapted from the U.S. Geological Survey (http://pubs.usgs.gov/fs/2007/3015/). Red lines indicate major unconformities between eras. Included are major events in life history and tectonic events occurring on the North American continent. Absolute ages shown are in millions of years (Ma).

Eon	Era	Period	Epoch	Events
Phanerozoic	Cenozoic	Quaternary	Holocene	18 Ka: Chesapeake Bay forms, shorelines evolve
			Pleistocene	Dramatic climate oscillations, rise & fall of sea-level cutting scarps along major rivers
		Tertiary	Pliocene	Marine sedimentation
			Miocene	Chesapeake group, erosional interval
			Oligocene	Erosional interval
			Eocene	35.7 Ma: Chesapeake Bay Impact Structure
			Paleocene	Erosional interval
	Mesozoic	Cretaceous		Shallow sea covers eastern Virginia
		Jurassic		Atlantic Ocean opens, East flowing rivers develop
		Triassic		Atlantic rifting begins- Deposition of sediments in rift basins
	Paleozoic	Permian		325-265 Ma: ALLEGHANIAN OROGENY
		Pennsylvanian		Coals deposited in coastal swamps 300 Ma: Petersburg granite emplaced
		Mississippian		Passive margin sedimentation
		Devonian		360 Ma: ACADIAN OROGENY
		Silurian		Taconic highlands eroded
		Ordovician		440-420 Ma: TACONIC OROGENY
		Cambrian		Carbonate deposition on passive margin
Proterozoic	Neoproterozoic			600-550 Ma: Late phase of Iapetan rifting
				750-700 Ma: Early phase of Iapetan rifting
	Mesoproterozoic			1100-950 Ma: GRENVILLIAN OROGENY
	Paleoproterozoic			

Figure 9. Geologic time scale specific to Virginia. Dates are approximate (Ma: millions of years ago; Ka: thousands of years ago). Rocks mapped within Appomattox Court House National Historical Park are Neoproterozoic and Cambrian in age. Graphic adapted from Bailey and Roberts (no date) (College of William & Mary).

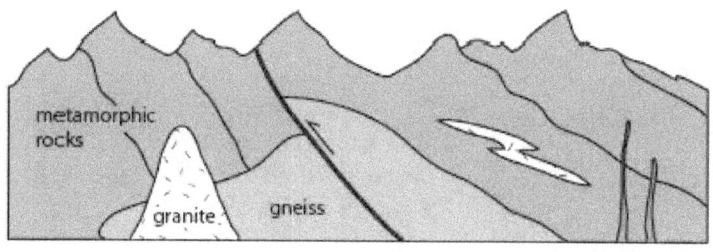

A) Middle Proterozoic, 1000 Ma
Granite gneisses form as a result
of compressive forces of Grenville
Orogeny, proto-Appalachian Mtns.

Erosion bevels the proto-Appalachian
highland and igneous activity
begins associated with extensional
tectonics

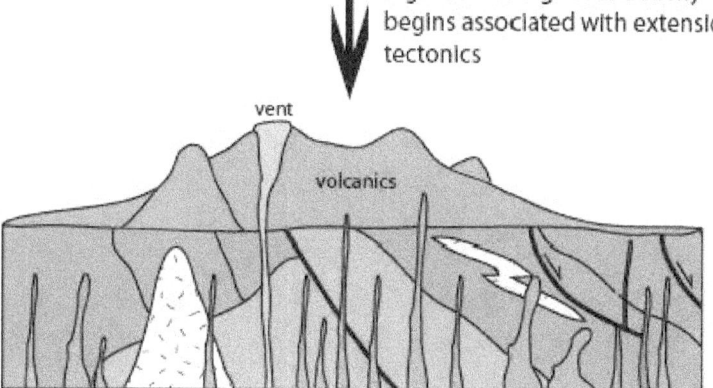

B) Late Proterozoic, 770–575 Ma
Catoctin Greenstone forms from
lava flows and volcanism during
continental rifting, Iapetus Ocean

Oceanic transgression creates
deposits of sands, muds and
carbonate atop the eroded
volcanic rocks

C) Cambrian, 545 Ma
Fossils appear, continental margin
and shelf develop

Figure 10. Geologic evolution of the Appalachian Mountains in the Appomattox area, west to east cross-sectional view. First, intrusions of granitic gneiss, metamorphism, and deformation related to the Grenvillian Orogeny lasted 60 million years, from 1.1 billion to 950 million years ago; these rocks are found in the Blue Ridge province (A). Then, continental rifting and volcanic activity in the Grenville terrane and turbidites deposited in the deep water basin to the east; this activity continued for about 200 million years, from about 770 to 575 million years ago (B). Next, the margin of the continent became stable with carbonate rocks deposited in quiet water (rocks of the current Great Valley). Shelly fossils appeared about 545 million years ago; then deep water rocks were deposited into a basin east of the shelf margin for about 65 million years (C). Adapted from Southworth and others (2001).

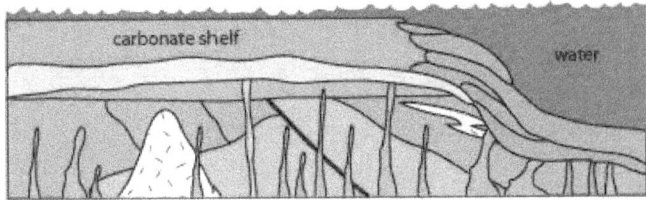

A) Cambrian and Ordovician Carbonate shelf thickens, platform edge, and oceanic basin develop on passive margin of continent

Compression from the east begins to deform and uplift continental margin. Oceanic crust and sediments thrust onto margin.

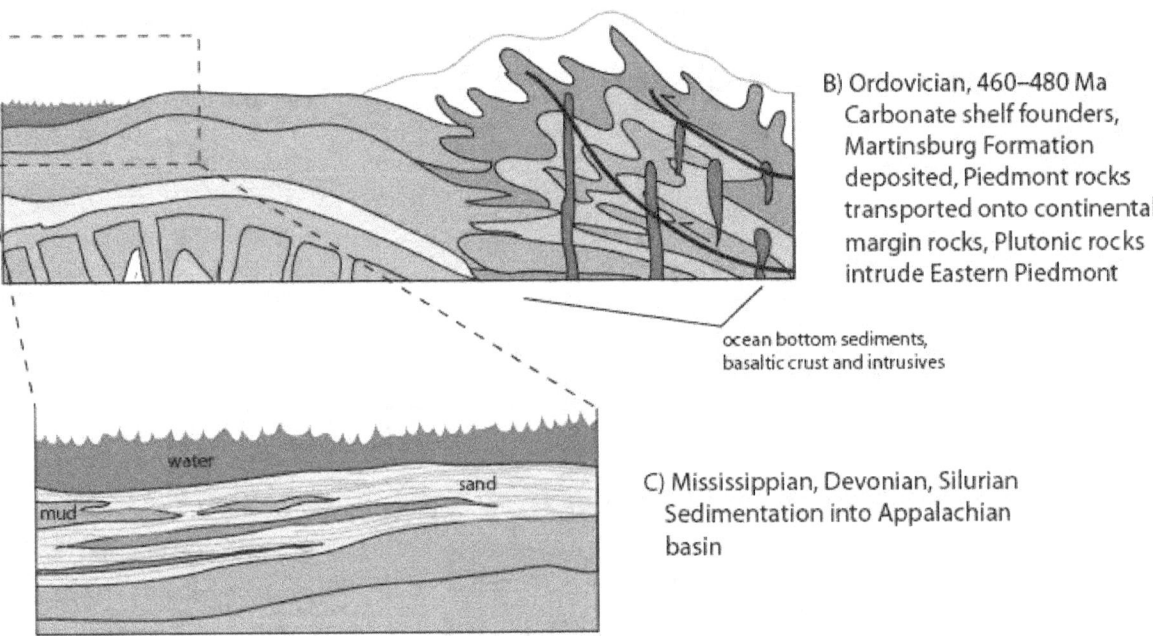

B) Ordovician, 460–480 Ma Carbonate shelf founders, Martinsburg Formation deposited, Piedmont rocks transported onto continental margin rocks, Plutonic rocks intrude Eastern Piedmont

ocean bottom sediments, basaltic crust and intrusives

C) Mississippian, Devonian, Silurian Sedimentation into Appalachian basin

Following deposition in the Appalachian Basin, compressional tectonics begins to fold and buckle sedimentary rocks and thrust oceanic crust and sediments onto eastern margin of North American continent

Figure 11. Geologic evolution of the Appalachian Mountains in the Appomattox area, west to east cross-sectional view. Following deposition, the stable shelf foundered as the Taconic Orogeny (480-460 million years ago) elevated the rocks to the east and provided a source for the clastic materials in Ordovician shales (A). Rocks in the Piedmont province were intruded by plutonic rocks (B). Then, a thick sequence of sedimentary rocks were deposited in a deepening Appalachian basin for 120 million years (C). Most of these rocks are now found in the Valley and Ridge province. Adapted from Southworth and others (2001).

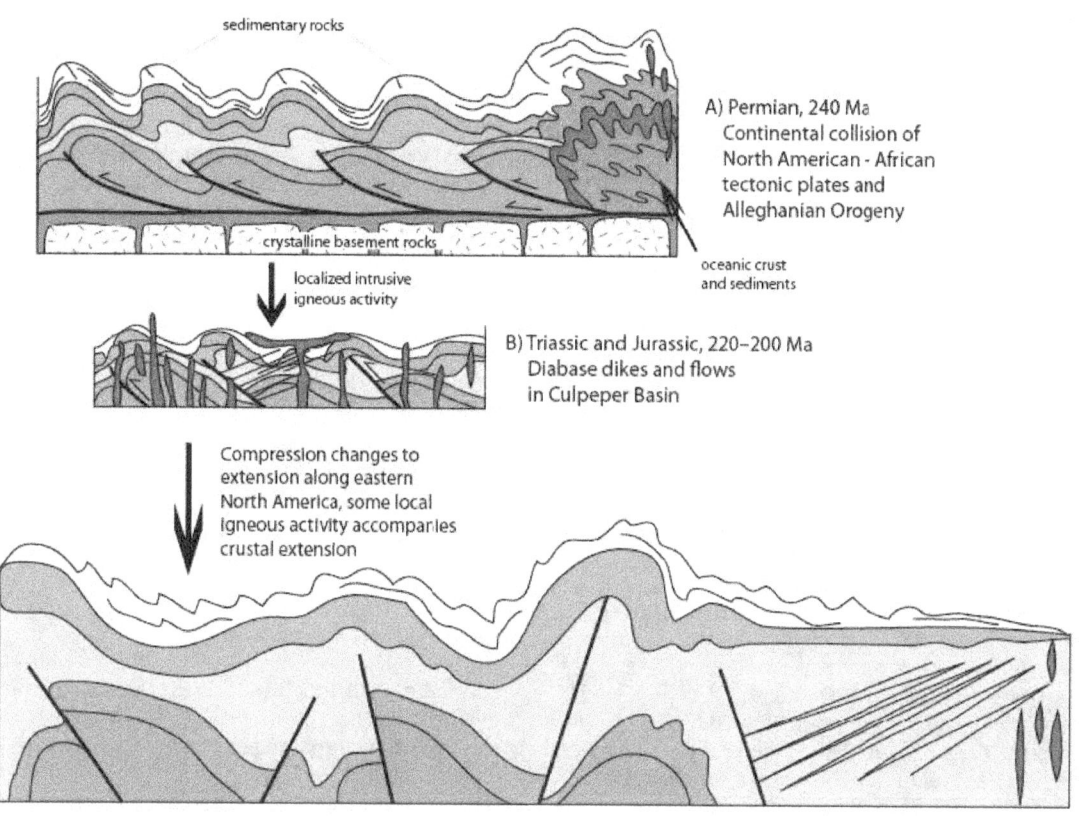

sedimentary rocks

A) Permian, 240 Ma
Continental collision of
North American - African
tectonic plates and
Alleghanian Orogeny

crystalline basement rocks

oceanic crust
and sediments

localized intrusive
igneous activity

B) Triassic and Jurassic, 220–200 Ma
Diabase dikes and flows
in Culpeper Basin

Compression changes to
extension along eastern
North America, some local
igneous activity accompanies
crustal extension

C) Cretaceous and Tertiary
Continental rifting creates basins and results in opening of Atlantic Ocean

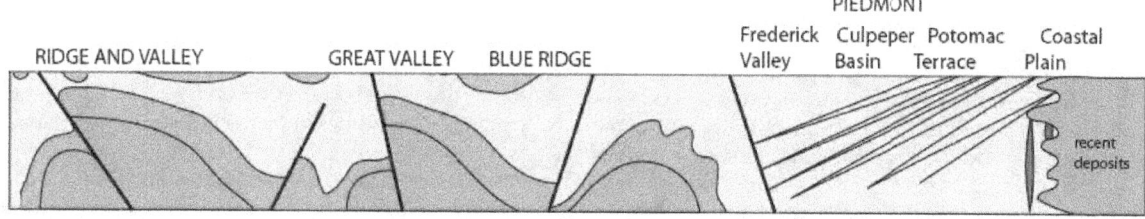

PIEDMONT

| RIDGE AND VALLEY | GREAT VALLEY | BLUE RIDGE | Frederick Valley | Culpeper Basin | Potomac Terrace | Coastal Plain |

recent deposits

D) Present
Erosion from highlands provides sediment deposited on Coastal Plain

Figure 12. Geologic evolution of the Appalachian Mountains in the Appomattox area, west to east cross-sectional view. About 240 million years ago, the continental tectonic plates of North America and Africa collided, resulting in the Alleghanian orogeny. Many of the folds and faults in rocks west of the Piedmont province are related to this event (A). About 20 million years later, continental rifting began and lasted for about 20 million years (220 to 200 million years ago) (B). Thick sequences of sedimentary rock were deposited in fault-bounded basins; volcanic activity occurred, ultimately resulting in the creation of the Atlantic Ocean. The Culpeper and Farmville basins in the western Piedmont are the result of this event (C). For the past 200 million years, the landscape has eroded, and rivers have carried the sediment eastward to deposit the thick strata of the Atlantic Coastal Plain (D). Diagrams are not to scale and are broadly representative of the tectonic settings. Adapted from Southworth and others (2001).

Glossary

This glossary contains brief definitions of technical geologic terms used in this report. Not all geologic terms used are referenced. For more detailed definitions or to find terms not listed here please visit: http://geomaps.wr.usgs.gov/parks/misc/glossarya.html.

accretion. The gradual addition of new land to old by the deposition of sediment or obduction of foreign landmasses onto a craton.

alluvial fan. A fan-shaped deposit of sediment that accumulates where a hydraulically confined stream flows to a hydraulically unconfined area. Commonly out of a mountain front into an area such as a valley or plain.

alluvium. Stream-deposited sediment.

amphibolite. A metamorphic rock consisting of amphibole and plagioclase with little to no quartz.

anticline. A fold, generally convex upward, with the stratigraphically older rocks in its core.

anticlinorium. A composite anticlinal structure of regional extent composed of lesser folds.

aphanitic. The texture of a fine-grained, igneous rock wherein the components are not distinguishable with the unaided eye.

aquifer. A rock or sedimentary unit that is sufficiently porous that it has a capacity to hold water, sufficiently permeable to allow water to move through it, and currently saturated to some level.

arc. See "island arc" and "volcanic arc."

ash (volcanic). Fine pyroclastic material ejected from a volcano (also see tuff).

barrier island. A long low narrow island formed by a ridge of sand that parallels the coast.

base level. The lowest level to which a stream can erode its channel. The ultimate base level for the land surface is sea level, but temporary base levels may exist locally.

basement. The undifferentiated rocks, commonly igneous and metamorphic, that underlie the rocks exposed at the surface.

basin (sedimentary). Any depression, from continental to local scales, into which sediments are deposited.

basin (structural). A doubly plunging syncline in which rocks dip inward from all sides.

beach. A gently sloping shoreline covered with sediment, commonly formed by the action of waves and tides.

bed. The smallest sedimentary strata unit, commonly ranging in thickness from one centimeter to a meter or two and distinguishable from beds above and below.

bedding. Depositional layering or stratification of sediments.

bedrock. The underlying solid rock as it would appear with the sediment, soil, and vegetative cover stripped away.

block (fault). A crustal unit bounded by faults, either completely or in part.

breccia. A coarse-grained, generally unsorted sedimentary rock consisting of cemented angular clasts greater than 2 mm (0.08 in).

calcareous. Describes rock or sediment that contains calcium carbonate.

carbonate. A mineral that has CO_3^{-2} as its essential component (e.g., calcite and aragonite).

cataclasite. A rock produced by the action of severe mechanical stress in a brittle regime, often containing angular clasts in a finer grained matrix.

chemical sediment. A sediment precipitated directly from solution (also called nonclastic).

chemical weathering. Chemical breakdown of minerals at the Earth's surface via reaction with water, air, or dissolved substances; commonly results in a change in chemical composition more stable in the current environment.

clastic. Rock or sediment made of fragments or pre-existing rocks.

clay. Can be used to refer to clay minerals or as a sedimentary fragment size classification (less than 1/256 mm [0.00015 in]).

conglomerate. A coarse-grained, generally unsorted, sedimentary rock consisting of cemented rounded clasts larger than 2 mm (0.08 in).

continental crust. The crustal rocks rich in silica and alumina that underlie the continents; ranging in thickness from 35 km (22 mi) to 60 km (37 mi) under mountain ranges.

continental shelf. The shallowly submerged portion of a continental margin extending from the shoreline to the continental slope with water depths of less than 200 m (656 ft).

convergent boundary. An active boundary where two tectonic plates are colliding.

craton. The relatively old and geologically stable interior of a continent.

cross-bedding. Uniform to highly varied sets of inclined sedimentary beds deposited by wind or water that indicate distinctive flow conditions (e.g., direction and depth).

cross-section. A graphical interpretation of geology, structure, and/or stratigraphy in the third (vertical) dimension based on mapped and measured geological extents and attitudes depicted in a vertically oriented plane.

crust. The Earth's outermost compositional shell, 10 to 40 km (6 to 25 mi) thick, consisting predominantly of relatively low-density silicate minerals (also see "oceanic crust" and "continental crust").

crystalline. Describes a regular, orderly, repeating geometric structural arrangement of atoms.

debris flow. A moving mass of rock fragments, soil, and mud, more than half the particles of which are larger than sand size.

deformation. A general term for the process of faulting, folding, and shearing of rocks as a result of various Earth forces such as compression (pushing together) and extension (pulling apart).

delta. A sediment wedge deposited where a stream flows into a lake or sea.

dextral. Pertaining to the clockwise direction. Faults are said to be "dextral" if relative movement across the fault is to the right. Opposite of "sinistral."

diabase. An intrusive igneous rock consisting of labradorite and pyroxene.

dike. A tabular discordant igneous intrusion.

dip. The angle between a bed or other geologic surface and horizontal.

dip-slip fault. A fault with measurable offset where the relative movement is parallel to the dip of the fault.

drainage basin. The total area from which a stream system receives or drains precipitation runoff.

eustatic. Relates to a simultaneous worldwide rise or fall of sea levels in the Earth's oceans.

extrusive. Of or pertaining to the eruption of igneous material onto the Earth's surface.

facies (metamorphic). The pressure-temperature regime that results in a particular distinctive metamorphic mineralogy (i.e., a suite of index minerals).

fault. A break in rock along which relative movement has occurred between the two sides.

felsic. An igneous rock having abundant light-colored minerals such as quartz, feldspars, or muscovite. Compare to "mafic".

felsite. A general term for any light-colored aphanitic igneous rock composed chiefly of quartz and feldspar.

foliation. A preferred arrangement of crystal planes in minerals; in metamorphic rocks, the term commonly refers to a parallel orientation of planar minerals such as micas.

formation. Fundamental rock-stratigraphic unit that is mappable, lithologically distinct from adjoining strata, and has definable upper and lower contacts.

fracture. Irregular breakage of a mineral. Any break in a rock (e.g., crack, joint, fault).

fumarole. A hole or vent from which volcanic fumes or vapors issue.

gabbro. A group of dark-colored, intrusive igneous rocks composed principally of calcic plagioclase and clinopyroxene. The coarse-grained equivalent of basalt.

gamma log. The radioactivity curve of the intensity of broad-spectrum undifferentiated natural gamma radiation emitted from the rocks in a cased or uncased borehole.

gneiss. A foliated rock formed by regional metamorphism with alternating bands of dark and light minerals.

graben. A down-dropped structural block bounded by steeply dipping, normal faults (also see "horst").

granodiorite. A group of coarse-grained plutonic rocks intermediate in silica composition containing quartz, oligoclase or andesine, and potassium feldspar.

greenstone. A field term for any compact dark green altered or metamorphosed basic igneous rock owing its color to chlorite, actinolite, or epidote.

horst. Areas of relative up between grabens, representing the geologic surface left behind as grabens drop. The best example is the basin and range province of Nevada. The basins are grabens and the ranges are weathered horsts. Grabens become a locus for sedimentary deposition.

hydraulic conductivity. Measure of permeability coefficient.

igneous. Refers to a rock or mineral that originated from molten material; one of the three main classes of rocks—igneous, metamorphic, and sedimentary.

intrusion. A body of igneous rock that invades (pushes into) older rock. The invading rock may be a plastic solid or magma.

island arc. A line or arc of volcanic islands formed over and parallel to a subduction zone.

isoclinal. Characteristic of a fold with limbs that are parallel.

isostasy. The process by which the crust "floats" at an elevation compatible with the density and thickness of the crustal rocks relative to underlying mantle.

isostatic adjustment. The shift of the lithosphere to maintain equilibrium between units of varying mass and density; excess mass above is balanced by a deficit of density below, and vice versa.

joint. A semi-planar break in rock without relative movement of rocks on either side of the fracture surface.

kaolinite. A common clay mineral of the kaolin group with a high alumina content and white color.

karst topography. Topography characterized by abundant sinkholes and caverns formed by the dissolution of calcareous rocks.

lacustrine. Pertaining to, produced by, or inhabiting a lake or lakes.

landslide. Any process or landform resulting from rapid, gravity-driven mass movement.

lava. Still-molten or solidified magma that has been extruded onto the Earth's surface though a volcano or fissure.

lineament. Any relatively straight surface feature that can be identified via observation, mapping, or remote sensing, often reflects crustal structure.

lithology. The physical description or classification of a rock or rock unit based on characters such as its color, mineralogic composition, and grain size.

lithosphere. The relatively rigid outmost shell of the Earth's structure, 50 to 100 km (31 to 62 miles) thick, that encompasses the crust and uppermost mantle.

mafic. Describes dark-colored rock, magma, or minerals rich in magnesium and iron.

magma. Molten rock beneath the Earth's surface capable of intrusion and extrusion.

mantle. The zone of the Earth's interior between crust and core.

massif. Large topographical feature formed as more rigid rocks protrude above the softer surrounding rock. Associated with orogenic areas.

matrix. The fine grained material between coarse (larger) grains in igneous rocks or poorly sorted clastic sediments or rocks. Also refers to rock or sediment in which a fossil is embedded.

meanders. Sinuous lateral curve or bend in a stream channel.

mélange. A mappable body of rock that includes fragments and blocks of all sizes, embedded in a fragmented and generally sheared matrix.

meta–. A prefix used with the name of a sedimentary or igneous rock, indicating that the rock has been metamorphosed.

metagraywacke. Metamorphosed remnant of an assortment of poorly sorted angular to subangular grains of quartz and feldspar with various dark mineral fragments in a clay-rich matrix initially deposited as turbidity current.

metamorphism. Literally, a change in form. Metamorphism occurs in rocks through mineral alteration, genesis, and/or recrystallization from increased heat and pressure. Metamorphic rocks are one of three main classes of rocks—igneous, metamorphic, and sedimentary.

metarhyolite. Metamorphosed remnant of a silica-rich igneous extrusive rock with abundant quartz and alkali feldspar.

mid-ocean ridge. The continuous, generally submarine, seismic, median mountain range that marks the divergent tectonic margin(s) in the Earth's oceans.

migmatite. Literally, mixed rock, with both igneous and metamorphic characteristics due to partial melting during metamorphism.

mineral. A naturally occurring inorganic crystalline solid with a definite chemical composition or compositional range.

monzonite. A group of plutonic rocks containing approximately equal amounts of alkali feldspar and plagioclase, little or no quartz, and commonly augite as the main mafic mineral. Intrusive equivalent of latite.

mylonite. A metamorphosed deformed rock with a streaky or banded appearance produced by shearing under heat and pressure.

normal fault. A dip-slip fault in which the hanging wall moves down relative to the footwall.

obduction. The process by which the crust is thickened by thrust faulting at a convergent margin.

oceanic crust. The Earth's crust formed at spreading ridges that underlies the ocean basins. Oceanic crust is 6 to 7 km (3 to 4 miles) thick and generally of basaltic composition.

orogeny. A mountain-building event.

outcrop. Any part of a rock mass or formation that is exposed or "crops out" at the Earth's surface.

Pangaea. A theoretical, single supercontinent that existed during the Permian and Triassic periods.

parent (rock). The original rock from which a metamorphic rock was formed. Can also refer to the rock from which a soil was formed.

passive margin. A margin where no plate-scale tectonism is taking place; plates are not converging, diverging, or sliding past one another. An example is the east coast of North America. (also see "active margin").

pebble. Generally, small rounded rock particles from 4 to 64 mm (0.16 to 2.5 in) in diameter.

peridotite. A coarse-grained plutonic rock composed chiefly of olivine and other mafic minerals; commonly alters to serpentinite.

permeability. A measure of the relative ease with which fluids move through the pore spaces of rocks or sediments.

phenocrysts. A coarse crystal in a porphyritic igneous rock.

phyllite. A metamorphosed rock with a silky sheen, intermediate in composition between slate and mica schist.

plateau. A broad flat-topped topographic high of great extent and elevation above the surrounding plains, canyons, or valleys (both land and marine landforms).

pluton. A body of intrusive igneous rock.

plutonic. Igneous rock intruded and crystallized at some depth in the Earth.

porphyritic. An igneous rock characteristic wherein the rock contains conspicuously large crystals in a fine-grained groundmass.

porphyroblastic. Describes a metamorphic rock with large crystals (such as garnets) in a much finer-grained (smaller crystals) matrix.

pyroxenite. An ultramafic plutonic rock chiefly composed of pyroxene with accessory mafic minerals.

recharge. Infiltration processes that replenish groundwater.

regolith. General term for the layer of rock debris, organic matter, and soil that commonly forms the land surface and overlies most bedrock.

regression. A long-term seaward retreat of the shoreline or relative fall of sea level.

reverse fault. A contractional high-angle (greater than 45°) dip-slip fault in which the hanging wall moves up relative to the footwall (also see "thrust fault").

rift valley. A depression formed by grabens along the crest of an oceanic spreading ridge or in a continental rift zone.

sandstone. Clastic sedimentary rock of predominantly sand-sized grains.

scarp. A steep cliff or topographic step resulting from displacement on a fault, or by mass movement, or erosion.

sediment. An eroded and deposited, unconsolidated accumulation of rock and mineral fragments.

sedimentary rock. A consolidated and lithified rock consisting of clastic and/or chemical sediment(s). One of the three main classes of rocks—igneous, metamorphic, and sedimentary.

schist. A strongly foliated crystalline metamorphic rock with parallelism of more than 50 percent of the component minerals, including abundant micas.

shale. A clastic sedimentary rock made of clay-sized particles that exhibit parallel splitting properties.

shear zone. A tabular zone of rock that has been crushed and brecciated by many parallel fractures due to shear strain.

sill. A tabular, igneous intrusion that is concordant with the country rock.

silt. Clastic sedimentary material intermediate in size between fine-grained sand and coarse clay (1/256 to 1/16 mm [0.00015 to 0.002 in]).

sinistral. Pertaining to the counterclockwise direction. Faults are said to be "sinistral" if relative movement across the fault is to the left. Opposite of "dextral."

siltstone. A variably lithified sedimentary rock composed of silt-sized grains.

slope. The inclined surface of any geomorphic feature or rational measurement thereof. Synonymous with gradient.

slump. A generally large coherent mass movement with a concave-up failure surface and subsequent backward rotation relative to the slope.

soil. Surface accumulation of weathered rock and organic matter capable of supporting plant growth and often overlying the parent material from which it formed.

specific conductance. The measure of discharge of a water well per unit of drawdown.

spring. A site where water issues from the surface due to the intersection of the water table with the ground surface.

strata. Tabular or sheetlike masses or distinct layers of rock.

stratigraphy. The geologic study of the origin, occurrence, distribution, classification, correlation, and age of rock layers, especially sedimentary rocks.

stream. Any body of water moving under gravity flow and confined within a channel.

strike. The compass direction of the line of intersection of an inclined surface with a horizontal plane.

strike-slip fault. A fault with measurable offset where the relative movement is parallel to the strike of the fault. Said to be "sinistral" (left-lateral) if relative motion of the block opposite the observer appears to be to the left. "Dextral" (right-lateral) describes relative motion to the right.

subduction zone. A convergent plate boundary where oceanic lithosphere descends beneath a continental or oceanic plate and is carried down into the mantle.

suture. The linear zone where two continental landmasses become joined due to obduction.

syncline. A downward curving (concave up) fold with layers that dip inward; the core of the syncline contains the stratigraphically-younger rocks.

synclinorium. A composite synclinal structure of regional extent composed of lesser folds.

tectonics. The geological study of the broad structural architecture and deformational processes of the lithosphere and aesthenosphere.

terraces (stream). Step-like benches surrounding the present floodplain of a stream due to dissection of previous flood plain(s), stream bed(s), and/or valley floor(s).

terrane. A large region or group of rocks with similar geology, age, or structural style.

thrust fault. A contractional dip-slip fault with a shallowly dipping fault surface (less than 45°) where the hanging wall moves up and over relative to the footwall.

topography. The general morphology of the Earth's surface, including relief and locations of natural and anthropogenic features.

transpressional. A type of deformation in which there is strike-slip motion as well as some shortening (compression).

trend. The direction or azimuth of elongation or a linear geological feature.

tuff. Generally fine-grained igneous rock formed of consolidated volcanic ash.

uplift. A structurally high area in the crust, produced by movement that raises the rocks.

volcanic. Related to volcanoes. Igneous rock crystallized at or near the Earth's surface (e.g., lava).

volcanic arc. A commonly curved, linear, zone of volcanoes above a subduction zone.

water table. The upper surface of the saturated zone; the zone of rock in an aquifer saturated with water.

weathering. The set of physical, chemical, and biological processes by which rock is broken down.

References

This section lists references cited in this report as well as a general bibliography that may be of use to resource managers (indicated by an *). A more complete geologic bibliography is available from the National Park Service Geologic Resources Division.

Appomattox Historical Society. 2007. Welcome to Appomattox History! http://www.appomattoxhistory.org/ (accessed December 6, 2007).

Bailey, C. M. 1999. Generalized geologic terrane map of the Virginia Piedmont and Blue Ridge. College of William and Mary. http://web.wm.edu/geology/virginia/provinces/terranes.html (accessed June 6, 2005).

Bailey, C. M. 2000. Major faults and high-strain zones in Virginia. College of William and Mary. http://web.wm.edu/geology/virginia/provinces/faults.html (accessed October 29, 2007).

Bailey, C. M. and C. Roberts. no date. Geologic time chart of Virginia. College of William and Mary. http://web.wm.edu/geology/virginia/provinces/geotime.html (accessed October 29, 2007).

Bell, C. F., D. L. Belval, and J. P. Campbell. 1996. Trends in nutrients and suspended solids at Fall Line of five tributaries to the Chesapeake Bay in Virginia, July 1988 through June 1995. Water Resources Investigations 96-4191. Reston, VA: U.S. Geological Survey.

Belval, D. L. and J. P. Campbell. 1996. Water-quality data and estimated loads of selected constituents in five tributaries to the Chesapeake Bay at the Fall Line, Virginia, July 1993 through June 1995. Open-File Report 96-0220. Reston, VA: U.S. Geological Survey.

*Berquist, C.R., Jr. 2003. Digital Representation of the 1993 Geologic Map of Virginia – Expanded Explanation. Publication 174. Charlottesville, VA: Commonwealth of Virginia Department of Mines, Minerals, and Energy, Division of Mineral Resources.

Brown, W. R. and L. Pavlides. 1981. Melange terrane in the central Virginia Piedmont. Geological Society of America Abstracts with Programs 13 (7): 419.

Budke, E.H., Jr. 1992. The Dutch Gap fault system in the tri-cities area: preliminary results of a microgravity survey. Virginia Journal of Science. 43 (2): 262.

Dischinger, J. B., Jr. 1987. Late Mesozoic and Cenozoic stratigraphic and structural framework near Hopewell, Virginia. Scale 1:24,000. Bulletin 1567. Reston, VA: U.S. Geological Survey.

Drake, A. A., Jr., A. J. Froelich, R. E. Weems, and K. Y. Lee. 1994. Geologic map of the Manassas Quadrangle, Fairfax and Prince William counties, Virginia. Scale 1:24,000. Geologic Quadrangle Map GQ-1732. Reston, VA: U. S. Geological Survey.

Duffy, D. F. and G. R. Whittecar. 1991. Geomorphic development of segmented alluvial fans in the Shenandoah Valley, Stuarts Draft, Virginia. Geological Society of America Abstracts with Programs 23 (1): 24.

Fisher, G. W. 1976. The geologic evolution of the northeastern Piedmont of the Appalachians. Geological Society of America Abstracts with Programs 8 (2): 172-173.

Good, R. S. 1981. Geochemical exploration and sulfide mineralization. In Geologic investigations in the Willis Mountain and Andersonville quadrangles, Virginia. Anonymous, Virginia Division of Mineral Resources Publication 29: 48-69. Charlottesville, VA: Commonwealth of Virginia Department of Mines, Minerals, and Energy, Division of Mineral Resources.

Gottfried, D., A. J. Froelich, and J. N. Grossman. 1991. Geochemical data for Jurassic diabase associated with early Mesozoic basins in the Eastern United States: Farmville and Scottsville basins and vicinity, Virginia. Scale 1:125,000. Open-File Report 91-0322. Reston, VA: U.S. Geological Survey.

Hackley, P. C., J. D. Peper, W. C. Burton, and J. W. Horton, Jr. 2007. Northward extension of Carolina slate belt stratigraphy and structure, south-central Virginia: results from geologic mapping. American Journal of Science 307 (4): 749-771.

Harris, A. G., E. Tuttle, and S. D. Tuttle. 1997. Geology of National Parks. Dubuque, Iowa: Kendall/Hunt Publishing Company.

*Hilgard, E. W. 1891. Orange sand, Lagrange, and Appomattox. American Geologist 1891: 129-131.

Horton, J. W., Jr., J. N. Aleinikoff, W. C. Burton, J. D. Peper, and P. C. Hackley. 1999. Geologic framework of the Carolina slate belt in southern Virginia: insights from geologic mapping and U-Pb geochronology. Geological Society of America Abstracts with Programs 31 (7): 476.

Kauffman, M. E, and E. P. Frey. 1979. Antietam sandstone ridges: exhumed barrier islands or fault-bounded blocks? Abstracts with Programs - Geological Society of America 11 (1): 18.

Maccubbin, R. J. 1952. Heavy mineral studies of the sediments of the Appomattox River, Virginia. Virginia Journal of Science 3 (4): 330-331.

Marr, J. D., Jr. 1981. Stratigraphy and structure. In Geologic investigations in the Willis Mountain and Andersonville quadrangles, Virginia. Scale 1:62,500. Anonymous, Virginia Division of Mineral Resources Publication 29: 3-8. Charlottesville, VA: Commonwealth of Virginia Department of Mines, Minerals, and Energy, Division of Mineral Resources.

Marr, J. D., Jr. and P. C. Sweet. 1980. Geology of the Andersonville Quadrangle, Virginia. Scale 1:24,000. Virginia Division of Mineral Resources Publication 26: GM 104A. Charlottesville, VA: Commonwealth of Virginia Department of Mines, Minerals, and Energy, Division of Mineral Resources.

McFarland, E. R. 1997a. Controls on recharge to coastal plain aquifers along the Fall Zone in Virginia. Eos, Transactions, American Geophysical Union 78 (17): 165.

McFarland, E. R. 1997b. Hydrogeologic framework, analysis of ground-water flow, and relations to regional flow in the Fall Zone near Richmond, Virginia. Water Resources Investigations 97-4021. Reston, VA: U.S. Geological Survey.

Means, J. 1995. Maryland's Catoctin Mountain parks: an interpretive guide to Catoctin Mountain Park and Cunningham Falls State Park. Blacksburg, VA: McDonald & Woodward Publishing.

*Miller, A. J., J. A. Smith, and L. L. Shoemaker. 1984. Channel storage of fine-grained sediment in the Monocacy River basin. Eos, Transactions, American Geophysical Union 65 (45): 888.

Mixon, R. B., L. Pavlides, D. S. Powars, A. J. Froelich, R. E. Weems, J. S. Schindler, W. L. Newell, L E. Edwards, and L. W. Ward. 2000. Geologic Map of the Fredericksburg 30'x60' Quadrangle, Virginia and Maryland. Scale 1:100,000. Geologic Investigations Series Map I-2607. Reston, VA: U.S. Geological Survey.

National Park Service. 2007. Strategic plan for Appomattox Court House National Historical Park October 1, 2007 – September 30, 2011. http://www.nps.gov/apco/parkmgmt/upload/FY2007-2011%20Strategic%20Plan.pdf (accessed November 17, 2007).

Nelson, A. E., J. D. Marr, Jr., and K. E. Olinger. 1999. Reconnaissance bedrock geologic map of the Red House, Madisonville, Aspen, and Charlotte Court House 7.5-minute quadrangles, Charlotte, Prince Edward, Appomattox, Campbell, and Halifax counties, Virginia. Scale 1:48,000. Open-File Report 99-0538. Reston, VA: U.S. Geological Survey.

*Nickelsen, R. P. 1956. Geology of the Blue Ridge near Harpers Ferry, West Virginia. Geological Society of America Bulletin 67 (3): 239-269.

*Onasch, C. M. 1986. Structural and metamorphic evolution of a portion of the Blue Ridge in Maryland. Southeastern Geology 26 (4): 229-238.

Schwab, F. L. 1970. Origin of the Antietam Formation (late Precambrian?; lower Cambrian), central Virginia Journal of Sedimentary Petrology 40 (1): 354-366.

Simpson, E. L. 1991. An exhumed Lower Cambrian tidal-flat: the Antietam Formation, central Virginia, U.S.A, In Clastic tidal sedimentology, eds. D. G. Smith, B. A. Zaitlin, G. E. Reinson, and R. A. Rahmani. Memoir - Canadian Society of Petroleum Geologists 16: 123-133.

Southworth, S., D. K. Brezinski, R. C. Orndorff, P. G. Chirico, and K. M. Lagueux. 2001. Geology of the Chesapeake and Ohio Canal National Historical Park and Potomac River Corridor, District of Columbia, Maryland, West Virginia, and Virginia; A, Geologic map and GIS files (disc 1); B, Geologic report and figures (disc 2). Open-File Report 01-0188. Reston, VA: U. S. Geological Survey.

Spears, D. B., B. E. Owens, and C. M. Bailey. 2004. The Goochland-Chopawamsic Terrane Boundary, Central Virginia Piedmont. In Geology of the National Capital Region: field trip guidebook, eds. Scott Southworth and William Burton. U. S. Geological Survey Circular, C1264: 223-245. Reston, VA: U. S. Geological Survey.

Thomas, P. J., J. C. Baker, and T. W. Simpson. 1989a. Variability of the Cecil map unit in Appomattox County, Virginia. Soil Science of America Journal 53 (5): 1470-1474.

*Thomas, W. A., T. M. Chowns, D. L. Daniels, T. L. Neatherly, L. Glover, and R. J. Gleason. 1989b. The subsurface Appalachians beneath the Atlantic and Gulf coastal plains. In The Geology of North America F-2. Boulder, CO: Geological Society of America.

Topuz, E., G. P. Wilkes, J. R. Lucas, and D. LeVan. 1982. Methane resources of the unmineable coal seams in the Richmond Basin. Proceedings, Unconventional gas recovery symposium 1982: 135-143. Richardson, TX: Society of Petroleum Engineers.

Tuomey, M. 1842. Notice of the Appomattox coal pits. Farmers Register 10: 449-450.

U. S. Army Corps of Engineers. 1975. Flood plain information, James River, Appomattox County, Virginia: river mile 222 to river mile 279. Scale 1:12,500. VSWCB/IB-75/517.

Virginia Division of Mineral Resources. 2003. Digital representation of the 1993 geologic map of Virginia. Publication 174 [CD-ROM; 2003, December 31]. Richmond, VA: Commonwealth of Virginia Department of Mines, Minerals, and Energy. (Adapted from: Virginia Division of Mineral Resources. 1993. Geologic map of Virginia and Expanded Explanation. Scale 1:500,000. Richmond, VA: Commonwealth of Virginia Department of Mines, Minerals, and Energy.)

Whittecar, G. R. and D. F. Duffy. 2000. Geomorphology and stratigraphy of late Cenozoic alluvial fans, Augusta County, Virginia, U.S.A. In Regolith in the Central and Southern Appalachians, eds. G. M. Clark, H. H. Mills, and J. S. Kite. Southeastern Geology 39 (3-4): 259-279.

Woodson, B. R. and M. Afzal. 1976. The taxonomy and ecology of algae in the Appomattox River, Chesterfield County, Virginia. Virginia Journal of Science 27 (1): 5-9.

Zen, E-an. 1997a. The Seven-Storey River: Geomorphology of the Potomac River Channel Between Blockhouse Point, Maryland, and Georgetown, District of Columbia, With emphasis on the Gorge Complex Below Great Falls. Open-File Report 97-60. Reston, VA: U.S. Geological Survey.

Zen, E-an. 1997b. Channel Geometry and Strath Levels of the Potomac River Between Great Falls, Maryland and Hampshire, West Virginia. Open-File Report 97-480. Reston, VA: U.S. Geological Survey.

Appendix A: Geologic Map Graphic

The following page is a preview or snapshot of the geologic map for Appomattox Court House National Historical Park. For a poster-size PDF of this map or for digital geologic map data, please see the included CD or visit the Geologic Resources Inventory publications Web page
(http://www.nature.nps.gov/geology/inventory/gre_publications.cfm).

Geologic Map of Appomattox Court House National Historical Park

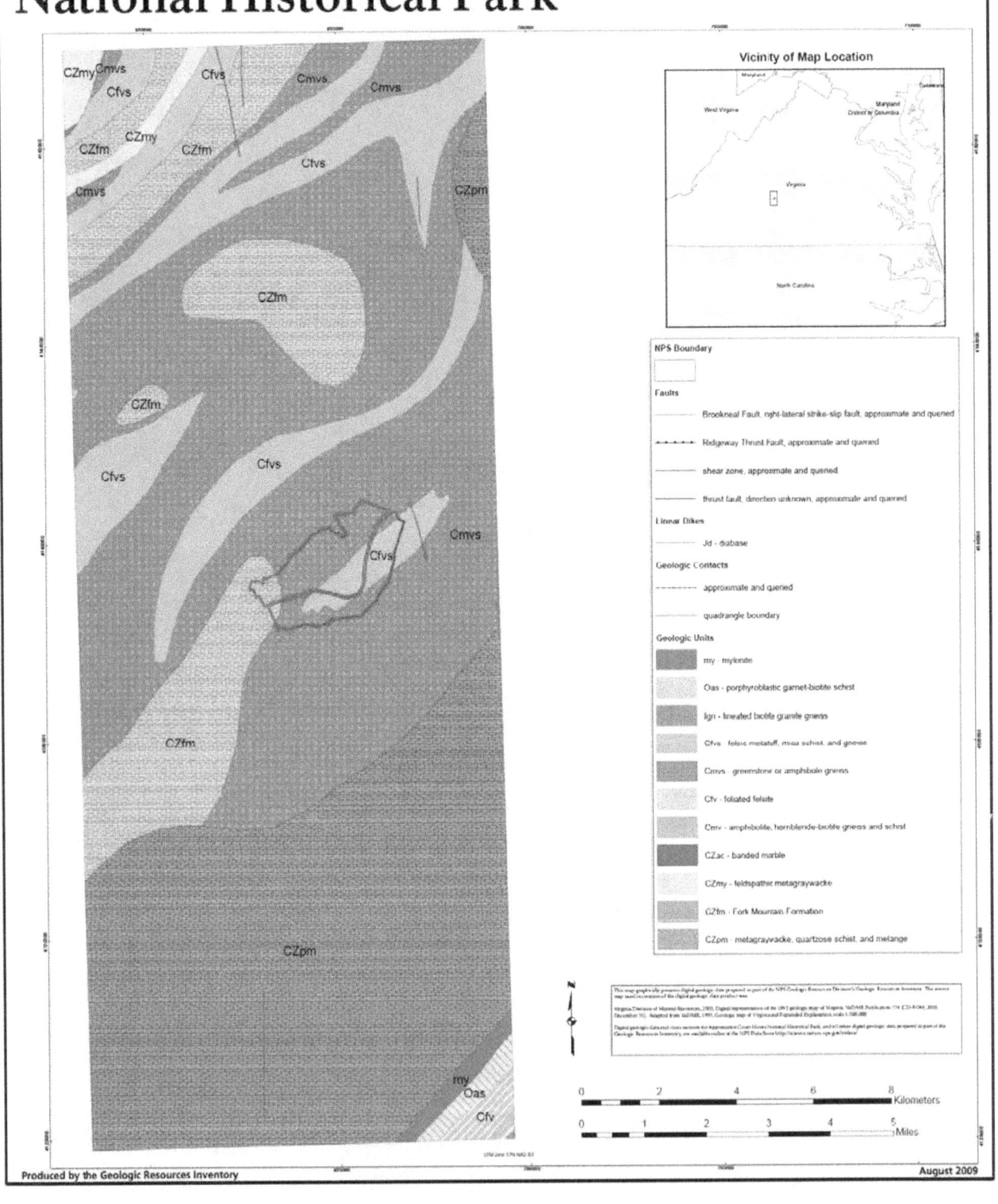

Vicinity of Map Location

NPS Boundary

Faults

———— Brookneal Fault, right-lateral strike-slip fault, approximate and queried

—▲—▲— Ridgeway Thrust Fault, approximate and queried

———— shear zone, approximate and queried

———— thrust fault, direction unknown, approximate and queried

Linear Dikes

———— Jd - diabase

Geologic Contacts

- - - - - approximate and queried

———— quadrangle boundary

Geologic Units

my - mylonite

Oas - porphyroblastic garnet-biotite schist

lgn - lineated biotite granite gneiss

Cfvs - felsic metatuff, mica schist, and gneiss

Cmvs - greenstone or amphibole gneiss

Cfv - foliated felsite

Cmv - amphibolite, hornblende-biotite gneiss and schist

CZac - banded marble

CZmy - feldspathic metagraywacke

CZfm - Fork Mountain Formation

CZpm - metagraywacke, quartzose schist, and melange

0		2		4		6		8	

Kilometers

0		1		2		3		4		5

Miles

Appendix B: Scoping Summary

The following excerpts are from the GRE scoping summary for Appomattox Court House National Historical Park. The scoping meeting was on April 21, 2005; therefore, the contact information and Web addresses referred to in this appendix may be outdated. Please contact the Geologic Resources Division for current information.

A Geologic Resources Evaluation scoping meeting for Appomattox Court House National Historical Park took place at Petersburg National Battlefield in Petersburg, VA on April 14, 2005. The scoping meeting participants identified the following list of geologic resource management issues. These topics are discussed in detail on pages 14-20.

1. Surface water quality at the headwaters of the Appomattox River is important to understand and monitor for the rest of the watershed

2. Seismic hazards

3. Erosion and slope processes

4. Geologic hazards including swelling soils

5. Connections between geology, biology, and the Civil War history to appeal to visitors interested in a deeper connection to the landscape

6. Human impacts including power lines, cattle grazing, nearby industry, increasing development

7. Mine-related features such as quarries, sand and gravel borrow pits

8. Hydrogeologic system characterization and groundwater quality to protect drinking supplies and understand how contaminants are moved through the system

9. Future facilities development and characterization with regards to geologic factors such as swelling clays and groundwater flow conduits

Appomattox Court House National Historical Park identified 2 quadrangles of interest, the Vera and Appomattox 7.5'x7.5' quadrangles. These quadrangles are all contained within the 30'x60' 1:100,000 Appomattox sheet. No known 1:1 map coverage for the quadrangles or sheet exists. However, the Virginia Division of Mineral Resources (VDMR) is interested in initiating some new mapping in the area as part of their 10 year STATEMAP program. The VDMR 1993, 1:500,000 scale map could be used in the interim, and additional information relevant to the area could be added to this coverage as reconnaissance mapping increases resolution of geologic map units. A definite time schedule has not been established.

Many other maps exist for the region that include coverage of the geology, oil and gas features, surficial geology, topography, groundwater features, land use,

Landsat imagery, geochemical features, aeromagnetic-gravity, mineral and mineral potential, hazard features, stratigraphy, hydrogeology, structures, glacial features, karst features, etc. The maps are available from agencies such as the U.S. Geological Survey, the Virginia Division of Mineral Resources, the Geological Society of America, the American Geological Institute, the Maryland Geological Survey, the West Virginia Geological and Economical Survey, and the West Virginia Geological Survey. Additional mapping at a smaller scale (1:6,000 ideally) within park boundaries will be more helpful for park resource management and interpretation.

Significant Geologic Resource Management Issues at Appomattox Court House National Historical Park
1. Surface water quality

One of the major goals of the park is to present the historical context of the area; this includes preserving and restoring any old buildings and the landscape around them. Maintaining this Civil War landscape often means resisting natural geologic changes, which presents several management challenges. Geologic slope processes such as chemical weathering, and slope erosion are constantly changing the landscape at the park. Alterations to park vegetation along exposed slopes lead to changes in the hydrologic regime at the park. For example, clearing of trees and their stabilizing roots for historical restoration, can lead to increased erosion, thereby increasing sediment load in nearby streams.

The river and other local streams are changing position constantly as part of natural meandering river flow. These shoreline changes threaten existing park facilities and the historical context of the landscape. Storm events including microbursts and thunderstorms, can send torrents of rain in a very localized area. Such storms are responsible for debris flows along slopes of the Blue Ridge Mountains and can cause slumping and landsliding on even moderate slopes.

Many waterways cross the landscape within and surrounding Appomattox Court House National Historical Park. These include the Appomattox River, North Branch, Plain Run Branch, and Rocky Run. A portion of the headwaters for the Appomattox River is contained within park boundaries. Because of this, the quality of the surface water at the park is very important to park management and by extension, the surrounding communities. Flooding and channel erosion are naturally occurring along most of the streams and rivers within the park.

Several riparian wetlands exist within the park and are threatened by increased flow and floods, as well as by sediment loading. Though small in scale at Appomattox Court House, wetlands are typically considered indicators of overall ecosystem health and should be researched and monitored periodically. Surface water quality at the park is threatened by ground compaction due to increased use as well as increases in impervious surfaces such as parking lots and roadways. These features increase the amounts of seasonal runoff as sheet flow.

Research and monitoring questions and suggestions include:
- Identify areas prone to slope failure during intense storm events.
- Monitor erosion rates and shoreline changes along the Appomattox River and compare to previous conditions using aerial photographs where available.
- What are the effects of increased erosion on aquatic ecosystems at the park?
- Is runoff in the park increasing due to surrounding development? If so, are there any remedial efforts the park can undertake to reduce this impact?
- Should the park preserve (recreate) historic landscapes at the expense of natural processes?
- Should the park target certain areas for restoration and leave others to natural processes?
- Develop an interpretive program detailing the balance between cultural context and natural processes at Appomattox Court House National Historical Park.
- Monitor water and soil quality in wetlands to establish as basis for comparison of future conditions.
- Use aerial photographs to study changes to wetland distribution through time.
- Determine any hotspots for water contamination. Remediate and monitor results.
- Research planting new vegetation along vulnerable reaches of park streams to prevent excess erosion and sediment loading.

2. Seismic hazards

Seismic events are not unheard of at Appomattox Court House National Historical Park. In 2003, a 4-4.8 magnitude earthquake occurred in the area. Possibly due to crustal relaxation (isostatic adjustment), earthquakes are not uncommon in the eastern United States. In addition to the ground shaking associated with earthquakes are landslides, damage to buildings and other manmade structures, ground and surface water disturbances, etc. Though the probability of a destructive seismic event at the park is low, resource management should be made aware of the potential.

Research and monitoring questions and suggestions include:

- Work with local universities and government agencies to monitor seismic activity in the area.
- Perform stability measurements of local slopes and historic building foundations to help predict possible responses to seismic events.

3. Erosion and slope processes

The topographic differences within and surrounding the historical park appear small and insignificant. However, the likelihood of landslides and slumps increases with precipitation and undercutting of slopes by streams, roads, trails, and other development in addition to natural erosion. Using a topographic map to determine the steepness of a slope, a geologic map to determine the rock type, and rainfall information, one could determine the relative potential (risk) for landslide occurrence.

Severe weather is difficult to predict. This area of Virginia can receive large amounts of snow and is in a hurricane-affected zone. Periodic flooding of the Appomattox River is the result of sudden, large inputs of precipitation. This extreme weather combined with moderate slopes, loose, unconsolidated soils and substrate can lead to sudden slope failures. In the vicinity of streams and rivers, this leads to shoreline erosion, increased sediment load, gullying, and threat of destruction for trails, bridges and other features of interest.

Increased erosion along the outer portions of bends in streambeds (where stream velocity is higher), causes the bank to retreat, undercutting the bank leading to washout. Trees, trails, and other features along these banks are undermined. Trees fall across the stream and trails are washed out. Remedial structures such as cribbing, log frame deflectors, jack dams, stone riprap, and/or log dams can shore up the bank, deflect the flow, and help to slow erosion (Means 1995).

Research and monitoring questions and suggestions include:
- Use shallow (10-inch) and deeper core data to monitor rates of sediment accumulation and erosion in the river, local streams, and springs.
- Monitor erosion rates by establishing key sites for repeat profile measurements to document rates of erosion or deposition, and reoccupy if possible shortly after major storm events. Repeat photography may be a useful tool.
- Perform a comprehensive study of the erosion and weathering processes active at the park, taking into account the different sediment deposit compositions versus slope aspects, location and likelihood of instability.
- Inventory runoff flood susceptible areas (paleoflood hydrology), relate to climate and confluence areas

4. Geologic hazards

Swelling soils occur in the Appomattox County area. Swelling soils (clays) expand when water saturated and shrink when dry. These soils are usually derived from weathered amphibolitic rocks, such as those present beneath the park. In extreme cases, these soils can undermine roads, buildings, and other structures. In less extreme ways, they can affect trails and other visitor use facilities.

Research and monitoring questions and suggestions include:
- Map locations of high swelling clay concentrations to avoid for future development.
- Perform a clay content survey to better understand soils and identify areas of swelling clay problems.

5. Connections between geology, biology, and the Civil War history

In many Civil War maneuvers, it was the advantage that familiarity with terrain, preparation to utilize of the natural features of the area, and the manipulation of the focal points, the gaps, ravines, cuts, hills, and ridges gave to one side or another that was to decide the outcome. The rolling hills and gentle landscape and topography at Appomattox Court House National Historical Park are defined by the geology and hydrology. This setting dictated the placement of the towns, strategy and encampment of troops, escape routes, river crossings, railroads, and the development of outlying areas. In addition to influencing battles, the landscape and topography also affected how troops and supplies were transported during the Civil War. Because Appomattox Court House is famous for the sight of Lee's surrender, geologic controls on the landscape and Civil War story are overlooked.

Runoff erodes sediments from any open areas and carries them down streams and gullies. Erosion naturally diminishes higher areas such as ridges and hills, foundations, earthworks, degrades bridge foundations, erodes streams back into restoration areas, and fills in the lower areas such as trenches, railroad cuts and stream ravines distorting the historical context of the landscape.

Interpreters make the landscape come alive for visitors and give the scenery a deeper meaning. Because geology forms the basis of the entire ecosystem, it is directly responsible for the history at Appomattox Court House National Historical Park. Thus, geologic features and processes should be emphasized to improve the visitor's experience.

The geology, in direct correlation with the soil types present at the park also controls natural biologic patterns in trees and other plants. The geologic units at Appomattox Court House National Historical Park include mafic metavolcanics and amphibolites. These calcium rich rocks weather to create fertile soils that attract specific native plants. For example, the American red cedar and redbuds prefer calcium rich substrates to grow in. The website for the park needs to be updated for geologic content and connections with other scientific (biology) and cultural disciplines.

Research and monitoring questions and suggestions include:
- Create interpretive programs concerning geologic features and processes and their effects on the Civil War history of the park.
- Perform and encourage outside research towards understanding geologic controls of plant distribution patterns within the park and surrounding areas.
- Encourage the interaction between geologists and the interpretive staff to come up with a list of features and programs to execute.
- Create a general interest map for visitors containing simple explanatory text on the geologic influences on troop movements.
- Update the park website relating geology with other resources.

6. Human impacts

The area surrounding Appomattox Court House National Historical Park is becoming increasingly populated. As development continues, conservation of any existing forest-meadow community types becomes a critical concern. Understanding the geology beneath the biotic communities becomes vital to their management. Park management of the landscape for historic preservation purposes compliments the preservation of these forests.

Humans began settling the Appomattox area in the 1700's, stemming from the earliest settlements further east in the Richmond area heading west towards the Blue Ridge. Their farming and homestead activities created an unnatural landscape that persists today at Appomattox Court House National Historical Park. Removal of soil and rocks, grazed pastures, and homestead sites dot the landscape.

Human impacts continue today as water lines, gas lines, power lines, radio towers, industrial complexes (along the southern boundary of the park), roads, and housing developments increase on the landscape. Cattle grazing on leased lands within and adjacent to the park threaten surface and groundwater quality as well as increases streambank erosion. Additionally, trails, visitor use areas, imported (invasive species), acid rain, and air and water pollution take their toll on the landscape. Resource management of these impacts is an ongoing process.

Research and monitoring questions and suggestions include:
- Perform acid rain measurements and correlate with the underlying bedrock to determine if any buffering effects occur. Relate this information to the water quality for the park.
- Are soils becoming more acidic due to acid rain?
- Monitor chemical alterations in bedrock.
- Cooperate with local developers and industry to minimize impact near park areas.

- Consult conservation groups regarding cooperative efforts to increase the areas of relevant parklands and protect more of the region around the park from development.
- Should streambanks affected by cattle grazing be remediated?
- Promote environmentally sound methods of developing land parcels including partial clearing of trees and proper construction of stable slopes.

7. Mine-related features

A historical depression within the park, now a vernal pool, may have been a borrow pit. Nearby stone quarries provided local building stones for Civilian Conservation Corps road building during the 1930's. Small-scale abandoned sand and gravel pits dot the landscape around Appomattox Court House National Historical Park. If historic, these features could be an interpretive program target. The state of Virginia has an inventory of mine features such as small open pit mines. Mines are constantly being identified and scanned into their database.

Research and monitoring questions and suggestions include:
- Comprehensively inventory mine-related features at Appomattox Court House National Historical Park, consulting aerial photographs and historic records if necessary.
- Consult the Virginia Division of Mineral Resources database for abandoned mine information.

8. Hydrogeologic system

The resource management needs to understand how water is moving through the hydrogeologic system into, under, and from the park. The geologic bedrock beneath this park is fractured and faulted. Such features provide quick conduits for water flow. Knowledge of the nature of the hydrogeologic system is critical to understanding the impacts of human induced contaminants on the ecosystem. The interaction between groundwater flow and the overall water quality should be quantitatively determined at the park. Little data regarding the nature of the hydrogeologic system at the park exists.

Park resource personnel also need to understand how the water table might change over time to manage water resources. Nearby cattle grazing threatens the groundwater quality in the area. There are several wells throughout the park that could be used for monitoring of ground water quality. It would be useful to perform tracer studies in these wells to see how quickly and in what direction water is moving through the system.

Research and monitoring questions and suggestions include:
- Inventory groundwater levels throughout the park.
- Test for and monitor organics (from agricultural and cattle waste), phosphate and volatile hydrocarbon

levels in the groundwater at the park, focusing on areas near facilities and at boundaries near industrial sites and housing developments.
- During detailed geologic mapping, inventory and map any existing springs in the park, especially with regards to their potential historical importance.
- Test water quality at any existing springs in the park.
- Create hydrogeologic models for the park to better manage the groundwater resource and predict the system's response to contamination.

9. Future facilities

Many geologic factors must be taken into account during decision making regarding siting future facilities. The development of visitor use sights including trails and picnic areas are now being considered by park management. High concentrations of shrink-and-swell clays could undermine any structure. Knowing the locations of springs and understanding the hydrogeologic system helps avoid problems with groundwater flow. Waste facilities should be sited taking hydrogeology into account as well as an understanding of the substrate.

Research and monitoring questions and suggestions include:
- Consult local geologic experts when planning future facilities at the park.
- Locate areas of high swelling clay concentrations, springs, groundwater flow, high slope, unstable substrate, etc. to avoid for future facilities.

Scoping Meeting Participants

Rick Berquist - Virginia Division of Mineral Resources, 757-221-2448, rick.berquist@dmme.virginia.gov and crberq@wm.edu

Mark Carter - Virginia Division of Mineral Resources, 434-951-6357, mark.carter@dmme.virginia.gov

Jim Comiskey - NPS, MIDN, 540-654-5328, jim_comiskey@nps.gov

Tim Connors - NPS, Geologic Resources Division, 303-969-2093, tim_connors@nps.gov

Brian Eick - NPS, APCO, brian_eick@nps.gov

Stephanie O'Meara - NPS, Natural Resources Information Division, 970-225-3584, Stephanie_O'Meara@partner.nps.gov

Trista Thornberry-Ehrlich - Colorado State University, 757-222-7639, tthorn@cnr.colostate.edu
Virginia Department of Mines, Minerals, and Energy, Division of Mineral Resources, Publication 174, 89 p.

Appomattox Court House National Historical Park
Geologic Resources Inventory Report

Natural Resource Report NPS/NRPC/GRD/NRR—2009/145

National Park Service
Acting Director • Dan Wenk

Natural Resource Stewardship and Science
Associate Director • Bert Frost

Natural Resource Program Center
The Natural Resource Program Center (NRPC) is the core of the NPS Natural Resource Stewardship and Science Directorate. The Center Director is located in Fort Collins, with staff located principally in Lakewood and Fort Collins, Colorado and in Washington, D.C. The NRPC has five divisions: Air Resources Division, Biological Resource Management Division, Environmental Quality Division, Geologic Resources Division, and Water Resources Division. NRPC also includes three offices: The Office of Education and Outreach, the Office of Inventory, Monitoring, and Evaluation, and Office of Natural Resource Information Systems. In addition, Natural Resource Web Management and Partnership Coordination are cross-cutting disciplines under the Center Director. The multidisciplinary staff of NRPC is dedicated to resolving park resource management challenges originating in and outside units of the National Park System.

Geologic Resources Division
Chief • Dave Steensen
Planning, Evaluation, and Permits Branch Chief • Carol McCoy
Geoscience and Restoration Branch Chief • Hal Pranger

Credits
Author • Trista Thornberry-Ehrlich
Review • Mark Carter and Jason Kenworthy
Editing • Bonnie Dash
Digital Map Production • Andrea Croskrey and Georgia Hybels
Map Layout Design • Andrea Croskrey, Georgia Hybels, and John Gilbert
Report Production • Lisa Fay

The Department of the Interior protects and manages the nation's natural resources and cultural heritage; provides scientific and other information about those resources; and honors its special responsibilities to American Indians, Alaska Natives, and affiliated Island Communities

NPS 340/100277, September 2009

www.ingramcontent.com/pod-product-compliance
Lightning Source LLC
Chambersburg PA
CBHW081624170526
45166CB00009B/3091